Self-Tutor for Computer Calculus Using *Mathematica*

D.C.M. Burbulla and C.T.J. Dodson
University of Toronto

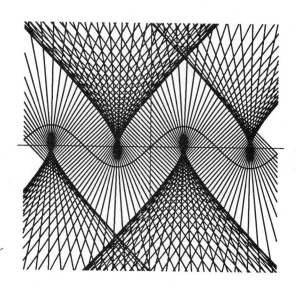

Canadian Cataloguing in Publication Data

Burbulla, D. C. M. (Dietrich C. M.), 1951-
    Self-tutor for computer calculus using mathematica

ISBN 0-13-803784-1

1. Calculus - Data processing. 2. Mathematica
(Computer program).   I. Dodson, C. T. J.   II. Title.

QA303.5.D37B8 1992   515'.0285   C92-093429-3

© 1992 Prentice-Hall Canada Inc., Scarborough, Ontario

ALL RIGHTS RESERVED

No part of this book may be reproduced in any form
without permission in writing from the publisher.

Prentice-Hall, Inc., Englewood Cliffs, New Jersey
Prentice-Hall International, Inc., London
Prentice-Hall of Australia, Pty., Ltd., Sydney
Prentice-Hall of India Pvt., Ltd., New Delhi
Prentice-Hall of Japan, Inc., Tokyo
Prentice-Hall of Southeast Asia (Pte.) Ltd., Singapore
Editora Prentice-Hall do Brasil Ltda., Rio de Janeiro
Prentice-Hall Hispanoamericana, S.A., Mexico

ISBN 0-13-803784-1

Acquisitions Editor: Jacqueline Wood
Developmental Editor: Judith Dawson
Production Coordinator: Lisa Kreuch
Cover Design: Olena Serbyn
Cover Images: Burbulla and Dodson using *Mathematica*

1 2 3 4 5   AP   96 95 94 93 92

Printed and bound in Canada by The Alger Press Limited

Contents

Preface   **xi**

1 Beginning *Mathematica*   **1**
   1.1 Getting Started 1
   1.2 On Screen 3
   1.3 Functions 5
   1.4 Symbols 10
   1.5 Lists and Tables 15
   1.6 Solving Equations 19
   1.7 Graphing 24
   1.8 Packages 36

2 Beginning Calculus   **41**
   2.1 Limits That *Mathematica* Can Do 41
   2.2 Limits That *Mathematica 1.2* Can't Do 46
   2.3 Continuity 54
   2.4 Optional: The Definition of a Limit 59

3 The Derivative   **67**
   3.1 The Derivative as Slope of the Tangent 67
   3.2 Differentiation with *Mathematica* 72
   3.3 Tangents and Normals 80
   3.4 Applications of the Derivative 94

4 Methods of Approximation   **103**

4.1	Linear Approximation	103
4.2	The Bisection Method	108
4.3	Newton's Method	114
4.4	Fixed Point Iteration	123

5 Properties of Differentiable Functions — 131

5.1	The Mean Value Theorem	131
5.2	Curvature	138

6 Antiderivatives — 151

6.1	Antidifferentiation	151
6.2	Constructing Antiderivatives	162

7 The Definite Integral — 175

7.1	Partitions and Riemann Sums	175
7.2	The Fundamental Theorem of Calculus	183

8 Numerical Integration — 193

8.1	The Trapezoid Rule and Simpson's Rule	193
8.2	Other Methods	202

9 Applications of the Integral — 209

9.1	Integral Formulas	209
9.2	Volume: Method of Cross Sections	210
9.3	Solid of Revolution	212
9.4	Volume: Method of Cylindrical Shells	213
9.5	Arc Length	214
9.6	Surface Area of Revolution	216
9.7	Separable Differential Equations	218

10 Exponential and Logarithmic Functions — 221

10.1	The Natural Logarithm as a Definite Integral	221
10.2	The Natural Exponential Function	223
10.3	Applications	226

11 Trigonometric and Hyperbolic Functions — **229**
- 11.1 Motivation 229
- 11.2 The Area of a Unit Circle 230
- 11.3 Derivatives and Integrals of Trigonometric Functions 231
- 11.4 Inverse Trigonometric Functions 232
- 11.5 Hyperbolic Functions 233

12 Techniques of Integration — **235**
- 12.1 Symbolic Integration 235
- 12.2 The Integration Game 243

13 Calculus Projects — **249**
- 13.1 Investigating Chaos 249
- 13.2 Newton's Method in the Complex Plane 259
- 13.3 Modelling 263

A *Mathematica* Functions Introduced — **267**
- A.1 Chap2.m 267
- A.2 Chap3.m 268
- A.3 Chap4.m 270
- A.4 Chap5.m 272
- A.5 Chap6.m 275
- A.6 Chap7.m 276
- A.7 Chap8.m 279
- A.8 Chap11.m 282
- A.9 Chap12.m 282
- A.10 Chap13.m 283

Bibliography — **285**

Index — **286**

List of Figures

1.1 Basic plotting with *Mathematica*: $\sin x$ on the left; $\sin x$ and x^2 on the right. 24

1.2 Four graphs plotted at once: $\sin x$, x^2, e^x, and e^{-x}. Which is which? . 25

1.3 The graph of $y = x^3 + 45x^2 - 574x - 4007$ on two different domains: on $[-4, 4]$ (top); and on $[-75, 25]$ (bottom). 26

1.4 The graph of $y = x^3 - 3x - 4$ to two different scales. 28

1.5 The graph of the circle $x^2 + y^2 = 9$ with two different aspect ratios; only with the option `AspectRatio -> Automatic` does the circle look like a circle. 30

1.6 Joining data points (top left) with straight line segments by using the option `PlotJoined -> True` (top right); fitting a curve to data points by using the function `Fit` (bottom). 31

2.1 Top: *Mathematica* plots $f(x) = \dfrac{x^3 - 1}{x^2 - 1}$ as if $f(1) = 1.5$. Bottom: *Mathematica* plots vertical lines in the graph of $g(x) = \dfrac{x^2 + 1}{1 - 3x^2}$ even though g is not defined at $x = \pm 1/\sqrt{3}$. 43

2.2 Top: $\lim\limits_{x \to 0} x \sin 1/x = 0$. Bottom: *Mathematica* gives a limit for the above function at $x = 0$, even though no limit exists. 48

2.3 Top: graph of $y = [[x]]$; no limit exists at $x = n$, for any integer n. Bottom: $\lim\limits_{x \to 0^-} e^{1/x} = 0$. 50

2.4 Top: graph of $y = \dfrac{|x - 3|}{x - 3}$ has a jump discontinuity at $x = 3$. Bottom: The graph of $y = \dfrac{x^2 + 1}{1 - x^2}$ has infinite discontinuities at $x = \pm 1$. 56

2.5 Plotting the graph of $y = x^3 + 4x^2 - 6$ to see how many real roots there are to the equation $x^3 + 4x^2 - 6 = 0$. 58

2.6 $\lim_{x\to 1} x^2 = 1$. For $\epsilon = 0.1$ four choices of δ are illustrated: $\delta = 0.1$ is too large (top left); $\delta = 0.05$ is slightly too large (top right); $\delta = 0.4$ (bottom left) or 0.3 (bottom right) are small enough. 61

2.7 $\lim_{x\to 1} x^2 = 1$. For $\epsilon = 0.01$, $\delta = 0.007$ is too large (left) but $\delta = 0.004$ is small enough. 61

2.8 Top left: $\lim_{x\to 0} x \sin 1/x \neq 1$. The limit is 0: if $\epsilon = 0.1$ take $\delta = 0.1$ (top right); if $\epsilon = 0.01, \delta = 0.1$ is too large (bottom right) but $\delta = 0.01$ is small enough (bottom left). 63

2.9 $h(x) = x$, if $x < 0$; $h(x) = x^2 + 1$, otherwise. Left: $\lim_{x\to 0} h(x) \neq 1$; right: nor is it 0. 63

2.10 Top: $\lim_{x\to 0+} h(x) = 1$. Bottom: $\lim_{x\to 0-} h(x) = 0$. 64

3.1 Top: family of secants to a curve passing through a common point. Bottom: normal and tangent to $y = \sin x$ at $x = 1$. 68

3.2 Functions which *Mathematica* cannot differentiate: it cannot differentiate $|x|$ (top left), $h(x)$ (top right), or $[[x]]$ (bottom right) anywhere. It can differentiate $x^2 \sin 1/x$ (bottom left) everywhere, except at $x = 0$. 74

3.3 Top: there are two tangents to the hyperbola $xy = 1$ passing through the point $(-1, 1)$. Bottom: there are three tangents to the cubic $y = 2x^3 + 13x^2 + 5x + 9$ passing through the point $(0, 0)$. 82

3.4 Top: the sum of the intercepts of any tangent to the graph of $\sqrt{x} + \sqrt{y} = \sqrt{k}$ is a constant. Bottom: tangent and normal to $y = x^2$ at the point $(-1, 1)$. 85

3.5 The family of tangents to the parabola with equation $y = x^2$; the curve is said to be the *envelope* of its tangents. 86

3.6 Families of normals to $y = x^2$ (top) and $y = \sin x$ (bottom). Note the interesting curves which form the envelopes of the normals. 88

3.7 The ellipse $4x^2 + 9y^2 = 45$ and the hyperbola $x^2 - 4y^2 = 5$ are orthogonal curves. 90

3.8 Shortest distance, from a point to a curve (top), and between curves (bottom), is measured perpendicularly. 91

3.9 Graphing a continuous function on a closed interval to see where its extreme values are. 94

3.10 The minimum of $a(x)$ is at the critical point; the maximum of $a(x)$ is at the right end point. 96

LIST OF FIGURES

3.11 Circle and square produced by cutting the wire at $x = 30$ (top) and $x = 45$ (bottom). 98

3.12 Circle and square produced by cutting the wire at $x = 70$ (top) and $x = 90$ (bottom). The maxmimum enclosed area is when the entire wire is bent into a circle. 99

3.13 As the ladder moves, the rates at which y and x change are related. . . 100

4.1 The tangent line to $y = \sqrt{x}$ at $x = 4$. 104

4.2 The absolute error of tangent line approximations to $y = \sqrt{x}$ on large intervals about $x = 4$. 106

4.3 The absolute error of tangent line approximations to $y = \sqrt{x}$ on small intervals about $x = 4$. 107

4.4 There are three real solutions to the equation $x^3 + 4x^2 - 6 = 0$ 109

4.5 Four applications of the Bisection Method. 110

4.6 Beginning Newton's Method. Top: initial guess 0.4; Bottom: after 2 iterations, approximation is 1.28603. 116

4.7 Further iterations of Newton's Method. Top: after 3 iterations approximation is 1.1062; Bottom: after 4 iterations approximation is 1.08636. 117

4.8 Beginning Fixed Point Iteration to solve the equation $\cos x = x$. . . . 125

4.9 Further iterations to find the fixed point of $\cos x$. 126

5.1 The Mean Value Theorem 132

5.2 The graphs of $y = 1 - 3x + x^3$ and its derivatives. 133

5.3 The graphs of $y = \dfrac{2x}{1+x^2}$ and its derivatives 134

5.4 The graphs of $y = x^{5/3} - 5x^{2/3}$ and its derivatives 136

5.5 Curvature is defined in terms of the angles that tangents make with respect to the x-axis 139

5.6 The curvature of $y = x^2$ 141

5.7 The osculating circle of $y = x^2$ at $x = 0$. 143

5.8 The osculating circle of $y = x^2$ at $x = 1$. 144

5.9 The evolute of $y = x^2$. 145

5.10 Osculating circles on different sides of an inflection point. 146

5.11 Graphs of $y = \sin x$, its curvature function, and its evolute. 147

5.12 Curvature details at two differnent points of $y = x^2$. 148

6.1 Antiderivatives of some well known functions: $\sin x$ (top); $\log x$ (bottom). A particular antiderivative must be specified by an initial condition. 157

6.2 Antiderivatives of some well known functions: x^2 (top); $\arctan x$ (bottom). A particular antiderivative must be specified by an initial condition. 158

6.3 The graph of $y = (1 + x^2)^{1/3}$. From its graph we can see that its antiderivatives must be increasing functions with an inflection point at $x = 0$. 162

6.4 The graph of $y = e^x \sin \pi x$, on $[0, 2]$. 165

6.5 Constructing an approximate antiderivative of $f(x) = e^x \sin \pi x$. Five points from f are used in the top left, 9 points are used in the top right, 17 points are used in the bottom left, 33 are used in the bottom right. 166

6.6 Approximate (top) and actual (bottom) antiderivatives of $f(x) = e^x \sin \pi x$. The approximation is based on 65 data points from the graph of f. 169

7.1 Approximating $\int_0^\pi \sin x\, dx$ with Riemann sums. Top: a regular partition of the interval; bottom: a random partition. 179

7.2 Approximating the area of a triangle with Riemann sums. 180

7.3 Top: the graph of $F(x) = \int_0^x t^2 dt$. Bottom: an antiderivative of $f(x) = x^2$. 184

7.4 Top: the graph of $F(x) = \int_0^x \cos(\cos t) dt$. Bottom: an antiderivative of $f(x) = \cos(\cos x)$. 186

7.5 Top: the graph of $F(x) = \int_{-2}^x h(t) dt$. Bottom: an antiderivative of $h(x)$. 187

7.6 Transforming a definite integral. The area of the region is the same before and after transformation. 189

8.1 Trapezoid rule approximations to $\int_0^1 \frac{4}{1+x^2} dx$ with $n = 4$ (top), and with $n = 10$, (bottom). 196

8.2 Simpson's rule approximations to $\int_0^1 \frac{4}{1+x^2} dx$ with $n = 4$ (top), and with $n = 10$, (bottom). 198

LIST OF FIGURES

8.3 Plots of $\frac{1}{|error|}$ against the number of intervals, n, for Riemann sum, trapezoid rule and Simpson's rule approximations. 199

8.4 Adaptive Numerical Integration. After adapting the partition to the shape of the curve the error is only 0.00434659. 205

9.1 Approximations to the sphere, cone and cylinder 211

9.2 Torus, formed by joining the ends of a cylinder or tube 213

9.3 Möbius strip, made from a flat strip joined at the ends after one twist; it has one edge and one surface 215

9.4 Some famous curves: the cardioid, the spiral of Archimedes and the limaçon 217

10.1 Plot of the integral $\int_1^x 1/t\, dt$ as a function of its upper limit 222

10.2 Plot of $exp\ x$ 223

10.3 Plot of $(1 + (1/n))^n$ 225

10.4 $(1 + (1/n))^n$ tends (slowly!) to e 226

10.5 Plot of the Poisson Distribution $e^{-2}\frac{2^n}{n!}$ 228

11.1 The areas of inscribed and exscribed n-gons for a unit circle tend to the limit π 231

13.1 $y = kx(1 - x)$; if $k > 0$ the parabola opens downwards; if $k < 0$ the parabola opens upwards. 250

13.2 Functional iteration applied to $y = kx(1 - x)$ with initial value 0.1. If $k = 2.6$ convergence to the stable equilibrium value of 0.615385 occurs (top); if $k = 3.2$ no convergence occurs after 101 iterations (bottom). 252

13.3 If $y = 4x(1 - x)$, functional iteration (applied to the initial choice of 0.1) exhibits no clear pattern. 254

13.4 For the graph of $g[3.2] \circ g[3.2]$, the two equilibrium values of $g[3.2]$ are both unstable. Successive iterations of $g[3.2]$ alternate between the two stable equilibrium values of $g[3.2] \circ g[3.2]$, namely 0.513045 and 0.799455. 255

13.5 The function $(1 + me^{-kt})^{-1/n}$ for $k = 1/2$, $m = 1$, $n = 1, 2, 3, 4, 5, 6$. . 264

13.6 The advance of perihelion of planetary orbits 265

Preface

This book is primarily a self-instructional companion to a computer-assisted first semester calculus course, like the one available at the University of Toronto, and can serve also as a self-tutor for anyone wishing to introduce themselves to *Mathematica* while reviewing some basic calculus. We have avoided repeating standard material found in calculus texts like Edwards and Penney [3] or Trim [7]. Our intention is that the user will have a designated course text as primary reference and employ this Self-Tutor in two ways:

- As an initial prop to help gain confidence quickly in using the computer as a study aid for new concepts in mathematics by exploiting graphics effectively

- Later as a source of ideas for exploring the theory and examples in the calculus course, as well as in all subsequent studies that involve mathematical concepts.

The software that we use is *Mathematica*[1], which comes with the definitive handbook by Wolfram [8]; this package is available on a variety of platforms and in student versions. *Mathematica* is a high-level language for doing mathematics, analytically as well as numerically, including statistics and data analysis, simulation and animation. It, or something very like it, seems destined to become an essential programming tool for the next generation of mathematicians and others who use mathematically based concepts. The most striking feature of *Mathematica* is its wide range of graphics capabilities. This facility, together with good on-line help and a mainly common-language based syntax, makes it easy for beginning students and others to use it for themselves with the minimum of introduction. So, very quickly they can investigate their own examples analytically, numerically, and graphically, then painlessly move on to animated graphics to portray connected *families* of examples. Since *Mathematica* knows all standard calculus, it also makes obsolete the tables of integrals, series and other functions. Moreover, its equation-solving capabilities meet most of the needs of those involved in mathematical modelling. For these reasons, our book may serve as a convenient self-tutor for anyone wishing to introduce themselves to *Mathematica* while reviewing some basic calculus.

---

[1] *Mathematica* is a registered trademark of Wolfram Research Inc.

The book is organized in a form to support active calculus laboratory sessions, whether directly supervised or not, with emphasis on self-study. The contents include notes on the topics cross-referenced to the detailed material in Edwards and Penney [3], ready-to-run analytical, numerical and graphical examples and further tutorial exercises to develop concepts. We have incorporated wherever possible, animations to see mathematics in motion, and maximum exploitation of graphics features. Animations can be used on DOS machines, Macintosh and NeXT workstations running Version 1.2 of *Mathematica* but Version 2 of *Mathematica*, which is beginning to be distributed, will extend animations to XTerminals. The student is advised to perform the whole sequence of *Mathematica* inputs and outputs in each section.

The first chapter gets the complete beginner started in *Mathematica* while reviewing some familiar material from high school mathematics. Subsequent chapters follow a more or less standard progression through the topics in a first semester calculus course for engineering and applied science students. However, since the idea of this Self–Tutor is to get students to the stage of using the computer to support their own study of the material in the course syllabus, we gradually withdraw and increasingly leave the student to investigate alone. The old adage, *I do, and I understand* is never truer than in mathematics! Seeing what you are doing by graphing functions adds enormously to interest and understanding.

We have presumed only basic familiarity with computer terminals, no experience with programming, and that students will not have easy access to expert humans. An Appendix gives a listing of functions in the *Mathematica* packages we introduce. These are also available on disk in DOS or UNIX format. Additionally, *Notebook* versions of the packages developed for this Self–Tutor are available now for NeXT computers and will be available for XWindows on UNIX platforms when the *Notebook* facility is available there. This book was prepared using LaTeX on a NeXT workstation with *PostScript*[2] graphics created in and then imported from *Mathematica*. The standard reference on LaTeX is Lamport [5]. There is now a growing number of books available using *Mathematica* for teaching, illustration and applications. In particular, a complete *Mathematica*–based calculus textbook has been written by Brown, Porta and Uhl [1]. More generally, for example, Maeder [6] shows the effectiveness of *Mathematica* as a programming language, Gray and Glynn [4] provide some fascinating glimpses of mathematics itself, Crandall [2] has provided some applications to the sciences and the *Mathematica Journal* published by Addison-Wesley contains information on current developments.

Dietrich Burbulla and Kit Dodson
Toronto, October 1991

---

[2] *PostScript* is a registered trademark of Adobe Systems Inc.

Chapter 1

Beginning Mathematica

> *The side of the square equals one. I have drawn four triangles in it. What is the surface area?*—Babylonian mathematics exercise in cuneiform Akkadian with graphics, on clay, circa 1700 BC

1.1 Getting Started

Mathematica runs on a variety of PCs and workstations, and under XWindows on some UNIX machines; this latter is the way you will probably use it most at first. Actually, any terminal (even via a modem) connected into a mainframe computer running *Mathematica* would allow you to do mathematical computations like algebra, differentiation and integration. However, you need a *graphics* terminal if you wish to take advantage of the graph plotting capabilities. The XTerminals or workstations that you will probably use in a calculus laboratory allow you to open one or more "windows", that is subscreens, for running different tasks simultaneously. For examle, in a typical *Mathematica* session you will have one window running the text input and output, another window will automatically be opened for any graph that you plot, and you may wish to use yet another window for copying output into a file that you create.

You may previously have used only PCs running under DOS, in which case the UNIX operating system may seem rather unforgiving at first. It is a good idea to prepare yourself by learning the basic commands, particularly for the editor `vi`, then practice opening and closing windows and transferring text between windows, and file saving. Once inside *Mathematica*, everything becomes easier. There, the query prefix, ?, is a very simple–to–use and friendly aid; for example, inputting `?Plot` will tell you how to use the *Mathematica* facility for plotting the graph of a function. You may find that you lose the graphics plots too easily, this can be eased by the following input line in *Mathematica*
`$ClickKillsWindow=False`

thereafter a graphics window will not go away until you point at it and type **q**.

You may have access to a PC on which you also run *Mathematica*, or which you use to prepare input for transferring to the UNIX server. In this case you need to read in from a floppy drive the DOS file, **myfundos.m** say, using **kermit** or a similar local file transfer system described in the local UNIX guide. Once your file is in your UNIX directory you can look at it with **vi**. You will probably find that it has lots of **M** characters, which DOS uses to command carriage return and a new line. You can remove these by the UNIX command

```
tr -s '\015' '\012' < myfundos.m > myfununix.m
```

While remaining inside *Mathematica*, you can execute any UNIX command by using the shell escape prefix, **!**. Thus, **!vi mary.m** calls up the vi editor to edit the file *mary.m*. You can leave **vi** in the usual way, **:wq**, returning to *Mathematica* at the point you left. Remember that UNIX, unlike DOS, is case sensitive — and so is *Mathematica*; this means that you must put capitals precisely where they appear in the definition of an operation or in the name of a file. It is unlikely that the coursework will result in you exceeding your memory allocation. However, graphics images can occupy a lot of space and you should familiarize yourself with the local commands to find out how much memory you are using. System administrators will usually provide increased file storage space for a legitimate purpose but you do need to ask in advance. There will also be some local rules about the use of printers, particularly laser printers; it is rather embarassing and expensive to cause large volumes of material to be printed out unnecessarily!

One of the advantages of networked systems is the electronic mail (email) facility, which allows you to send messages and files to other users. This is a good communication channel among fellow students, your TAs, and professors. Bear in mind, though, short enquiries are more likely to gain attention and elicit a quick response; email of the form: "Can you see what's wrong in ..." followed by several screens full of material is not a good idea without some prior arrangement! When you logon to the system you will be told of any email messages waiting for you and it is good practice to scan these immediately — there may be notice of a change in assignments or other information.

Finally, just as with a pencil and paper, doing mathematics by computer is more efficient in short sessions with breaks. If you are still stuck at the same point in a problem after half an hour, then you probably need some help. *Mathematica* quickly tells you if you are using one of its functions improperly and it can inform you about the available functions. Nevertheless, it cannot tell you which function you *need*, so do get into the habit of planning a teminal session beforehand by checking the definitions — in the mathematics and in the software. Even such a high level language as *Mathematica* obeys the Fundamental Theorem of Computing: garbage

1.2. ON SCREEN

in, garbage out! But used properly, you will quickly acquire a powerful investigative tool that can take most of the drudgery out of your mathematics, enabling you to study and visualize much more realistic situations in applications.

> **MATHEMATICA FUNCTIONS INTRODUCED IN THIS SECTION**
> ? !

1.2 On Screen

You should now be within reach of a terminal that has access to *Mathematica*, and in possession of any passwords and local commands needed to start it up—**math** and Enter starts *Mathematica*. Once started, the screen prompt should look something like:

`In[1]:=`

This simply tells you that it is awaiting your first input request. So give it something easy to understand in case things get out of hand. Try the following, noting the importance of spaces between characters, and of ending a command with the **Enter** key:

`In[1]:= 2 + 3`

The output comes automatically:

`Out[1]= 5`

`In[2]:=`

And again it awaits our request. Next try something a little bit more complicated, but still arithmetical, say $(2+3)^{23} - 3(6)^{14-2}$:

`In[2]:= (2 + 3)^23 - 3 6^(14 - 2)`
`Out[2]= 11920922424731117`

Note that round brackets are used for grouping, ^ is used for exponentiation, and a space between numbers can be used for multiplication. (Multiplication can also be represented by * .) Division is represented by /; try $21/46 - 13/67$.

```
In[3]:= 21/46 - 13/67
            809
Out[3]= ----
           3082
```

Note that *Mathematica* gives the exact answer, in rational form. If you want a decimal approximation you can add **// N** to your input:

```
In[4]:= 21/46 - 13/67 // N
Out[4]= 0.262492
```

Optionally, you can indicate to *Mathematica* that you wish an output in decimal form, simply by adding a decimal point in at least one number of your input. To refer to your last output you can use `%` , so that `In[4]` could have been typed as

```
In[4]:= % // N
Out[4]= 0.262492
```

Similarly, `%%` refers to the next to last output, `%%%` refers to the third to last output, and so on. If you know the actual output number, say n , then you can reference it with `%n`, or with `Out[n]`. (In a DOS environment you can also replace a current input with another by using the up-arrow or down-arrow keys, and you can scroll through your entire session by using the page-up or page-down keys; Macintosh and NeXT machines use the mouse.) Well known mathematical constants are represented in *Mathematica* by built in symbols, so that `Pi` represents π ; `E` represents e , the natural base of logarithms; and `I` represents i , the square root of minus one. For instance,

```
In[5]:= Pi // N
Out[5]= 3.14159
In[6]:= E // N
Out[6]= 2.71828
In[7]:= I^2
Out[7]= -1
```

To end a *Mathematica* session, type `Quit` and press `Enter`. To interrupt *Mathematica* while it is calculating, type Control-c — *Mathematica* will halt what it is doing and offer you a list of options, one of which is *abort*.

MATHEMATICA FUNCTIONS INTRODUCED IN THIS SECTION
+ - *
/ ^ %
N

Exercises:

1. Use *Mathematica* to find the following:

    (a) $1/5 + 2/3$

    (b) $\dfrac{1/3 - 4/7}{4/13 + 1/9}$

1.3. FUNCTIONS

     (c) 2^{40}
     (d) 2^{400}
     (e) 2^{4000}
     (f) $1/e + 1/\pi$
     (g) $e^\pi - \pi^e$
     (h) $e^{i\pi}$

2. Use *Mathematica* as a calculator, by adding `// N` to your inputs, to compute the answers of Exercise 1.

3. Enter `1/0` ; what does *Mathematica* return?

4. Enter `0/0` ; what does *Mathematica* return?

1.3 Functions

Mathematica offers a large selection of built-in functions. These are all represented by names beginning with a capital letter; the names are usually descriptive of what the function represents. The argument of a function in *Mathematica* is always put in square brackets, `[ ]`. Thus `Sin[x]`, `Log[x]`, `Sqrt[x]`, `Exp[x]` and `Abs[x]` evidently represent the functions

$$\sin x, \log x, \sqrt{x}, e^x \text{ and } |x|.$$

Note that, like mathematicians, *Mathematica* always uses `Log` to mean the *natural logarithm* function, the inverse of the natural exponential function, `Exp`. Some sample calculations:

```
In[1]:= Abs[-9]
Out[1]= 9
In[2]:= Sqrt[25]
Out[2]= 5
In[3]:= Sin[Pi/2]
Out[3]= 1
In[4]:= Log[E^2]
Out[4]= 2
In[5]:= Sin[Pi/3]
         Sqrt[3]
Out[5]= -------
           2
In[6]:= Log[234]
Out[6]= Log[234]
```

Observe that *Mathematica* gives exact answers, so that Out[5] and Out[6] are not approximated. If you want decimal approximations for these expressions you can of course use // N :

In[7]:= % // N
Out[7]= 5.45532

You can get on-line information about any function by entering ? followed by the function name.

In[8]:= ?Abs
Abs[z] gives the absolute value of the real or complex number z.

You can get more information, although it may be too technical to be helpful, by entering ?? followed by the function name. Try this for Abs, Sin and Log. By using the wild-card character * you can list more than one function:

In[8]:= ?N*

N	Nest	Null
NBernoulliB	NestList	NullSpace
NIntegrate	NonAssociative	NumValue
NProduct	NonCommutativeMultiply	Number
NRoots	NonConstants	NumberForm
NSum	NonNegative	NumberPoint
NameQ	None	NumberQ
Names	Normal	NumberSeparator
Needs	Not	Numerator
Negative		

Note that N is itself a fuction:

In[8]:= ?N
N[expr] gives the numerical value of expr. N[expr, n] does computations to
    n-digit precision.

So for instance $\sqrt{7}$ can be calculatd to 20 digit accuracy simply by:

In[8]:= N[Sqrt[7], 20]
Out[8]= 2.6457513110645905905

There are many built-in functions in *Mathematica*, but sometimes it will be necessary to define your own. (So as not to confuse your functions with those built in to *Mathematica*, it is a good idea to name yours in terms of lower case letters only.) This is easily done. For example, suppose you wish to define a function f such that $f(x) = x \log x$.

1.3. FUNCTIONS

```
In[9] := f[x_] := x Log[x]
```

You can check the definition of f by entering `?f`. *Mathematica* responds with:

```
f/:  f[x_] := x Log[x]
```

Now calculate $f(3)$, exactly and approximately, and $f(1 + a)$:

```
In[10] := f[3]
Out[10]= 3 Log[3]
In[11] := f[3] // N
Out[11]= 3.29584
In[12] := f[1 + a]
Out[12]= (1 + a) Log[1 + a]
```

In this example, the definition of f is straigtforward; it may happen, though, that you need to define a very complicated function — or that, you define a function which is a list of procedures (something like a program). In such cases, using the query prefix, ? will simply result in *Mathematica* printing the rules of a very complicated function, or a list of interconnected procedures — this information may cover pages, and may not be very helpful. In such cases you should make use of *Mathematica's* usage statements. These are invoked by including a second input of the form

```
f::usage = "... statement ... "
```

after the definition of f. As an example, working with $f(x) = x \log x$, here is a usage statement for f:

```
In[13]= f::usage = "f multiplies a positive number by its natural logarithm."
Out[13]= f multiplies a positive number by its natural logarithm.
```

Then if you query f with ?, *Mathematica* will respond with the usage statement only; however, if you query f with ??, *Mathematica* will respond with the usage statement followed by the actual definition of f:

```
In[14] := ?f
f multiplies a positive number by its natural logarithm.
In[14] := ??f
f multiplies a positive number by its natural logarithm.
f/:  f[x_] := x Log[x]
```

You will find that we have defined many functions for your use in later sections of this book, and in many cases we have supplied usage statements for the functions.

Unless specifically saved in a file (more details on how to do this later), any function that you define will be lost each time you end a *Mathematica* session. If during a session you wish to clear a function you have defined, enter `Clear[f]` .

There are at least four ways to evaluate a function at a given point, two of which we have already seen illustrated with `N` , namely `N[expr]` and `expr // N` These two ways, and two others, are illustrated for `Sin[Pi]` , which of course is equal to 0:

```
In[14]:= Sin[Pi]
Out[14]= 0
In[15]:= Pi // Sin
Out[15]= 0
In[16]:= Sin @ Pi
Out[16]= 0
In[17]:= Sin[x] /. x -> Pi
Out[17]= 0
```

Try these four ways on the function f defined above in `In[9]`; find $f(3)$.

You will be happy to know that *Mathematica* has functions which can do much of the calculus you have already learned (and has many additional functions to do things you have not learned yet!) Derivatives and indefinite integrals, to name but two things, can be typed into *Mathematica* by using the functions `D` and `Integrate` respectively.

```
In[18]:= D[Sin[x], x]
Out[18]= Cos[x]
In[19]:= Integrate[1/x, x]
Out[19]= Log[x]
```

An alternate way to differentiate a function is to use ' ; so for example `In[17]` could have been entered as `Sin'[x]` . Try differentiating and integrating $f(x)$, as defined above. Try it with pen and paper ... then try it with *Mathematica* :

```
In[20]:= D[f[x], x]    (* or In[19]:= f'[x] *)
Out[20]= 1 + Log[x]
In[21]:= Integrate[f[x], x]
                2       2
            -x      x  Log[x]
Out[21]=   --- +  ---------
             4         2
```

Note that remarks enclosed within parantheses with asterisks are ignored by *Mathematica*. This allows you to make notes along side your inputs, perhaps to remind

1.3. FUNCTIONS

you of something, perhaps for future reference, or perhaps, as in `In[20]` above, for the information of the reader.

The variety of the built-in functions in *Mathematica* is what gives it its great power. We will make a quick survey of some of the most commonly used of these functions in the next few sections, but it is revealing to point out now that essentially *everything* in *Mathematica* is a function. For instance, `+` , `<` , `=,` `==` , `If` are all examples of things that you may not have thought of as functions, but are correctly so represented in *Mathematica*.
(Try for yourself `?+` , `?<` , `?=` , `?==` , `?If`, `?Plot`.) So for instance:

```
In[22] := 5 < 4
Out[22]= False
In[23] := 2^3 == 8
Out[23]= True
```

Indeed, in every *Mathematica* dialogue, the inputs and outputs are in functional form: `In[n]` and `Out[n]`. It makes complete sense to speak of `In @ 4` and `7 // Out` .

```
In[24] := In @ 4    (* this will make input 24 the same as input 4 *)
Out[24]= 2
In[25] := 7 // Out
Out[25]= 5.45532
```

MATHEMATICA FUNCTIONS INTRODUCED IN THIS SECTION		
Sin	Cos	Tan
Csc	Sec	Cot
Exp	Log	Sqrt
Abs	D	Integrate
Clear	??	If

Remember, you can always review the definition of these functions with the query prefix: `?`.

Exercises:

1. Use *Mathematica* to calculate the following expressions to 10-digit precision.

    (a) $\sqrt{2}$

    (b) $\log 2$

    (c) e^{-1}

(d) $\dfrac{1+\sqrt{5}}{2}$

(e) $\sin \pi/12$

(f) $\tan(\pi/6)$

(g) $\arctan 1$

(h) $3^{1/5}$

2. Calculate π to 100-digit precision.

3. Question *Mathematica* about `Factorial`; calculate 9! .

4. Question *Mathematica* about `Prime` ; what is the 14*th* prime number?

5. Define a function g such that $g(x) = \sqrt{1+x^2}$. Calculate the following for g:

   (a) $g(2)$ to 6-digit precision

   (b) the approximate value of $g'(2)$

   (c) $g'(x)$

   (d) $\int g(x)dx$

6. Define a function h using the built-in function `If` such that
$$h(x) = x, \text{if} x < 0, \text{ but } h(x) = x^2 \text{ otherwise} .$$

   (a) Calculate $h(-3)$

   (b) Calculate $h(3)$

1.4 Symbols

With the help of functions such as `Expand`, `Factor` and `Simplify`, *Mathematica* can manipulate symbolic expressions.

```
In[1]:= (2 - x)^5
                5
Out[1]= (2 - x)
In[2]:= Expand[%]
                   2       3        4    5
Out[2]= 32 - 80 x + 80 x  - 40 x  + 10 x  - x
```

For fun you could now factor the latest result:

1.4. SYMBOLS

```
In[3]:= Factor[%]
              5
Out[3]= -(-2 + x)
```

Just as in written mathematics, it is possible in *Mathematica* to let a variable represent another expression. This is accomplished by using a single equal sign, = , between two expressions.

```
In[4]:= y = (1 + x^3)/(1 + x)
             3
         1 + x
Out[4]= ------
         1 + x
```

Any future reference to **y** will now assume it is equal to the above expression. Thus:

```
In[5]:= Simplify[y]
                 2
Out[5]= 1 - x + x
In[6]:= 1/y
         1 + x
Out[6]= ------
             3
         1 + x
```

Mathematica automatically updates the value of all other variables as soon as you assign any variable a new value. Thus if x is set equal to 2, then y will be set equal to 3.

```
In[7]:= x = 2
Out[7]= 2
In[8]:= y
Out[8]= 3
```

Try setting $x = 1 + a$; what is the value of y?

```
In[9]:= x = 1 + a
Out[9]= 1 + a
In[10]:= y
                    3
         1 + (1 + a)
Out[10]= ------------
             2 + a
```

```
In[11]:= Simplify[y]
                 2
Out[11]= 1 + a + a
```

If during a *Mathematica* session you forget the assigned value of a symbol, say `y`, you can simply enter `?y` and *Mathematica* will respond with its current value. Should you wish to clear a symbol of its assinged value, enter `Clear[y]`. A word of caution: before introducing a new variable it is always wise to check if it is already in use!

At this time please clear the variable `x` :

```
In[12]:= Clear[x]
```

There is another way to make symbolic substitutions without permanently reassigning a variable. For instance, with y as above we could find the value of y if x were set equal to $1 + a$ by entering `In[13]:= y /. x -> 1 + a` This gives the same output as `Out[10]` above, but does not set x equal to $1 + a$ for any future calculations. (That is, the effect of `->` is to make a *local* substitution, whereas the effect of `=` is to make a *global* substitution.) In general, replacements of the type *expression* `->` *new expression* can be extremely useful when trying to simplify complicated algebraic quantities. Now also clear the variable `y`:

```
In[14]:= Clear[y]
```

A string of characters with no spaces between is considered to be one symbol. Thus you can assign meaningful names to expressions, better to keep track of your work. For example:

```
In[15]:= quadratic = a x^2 + b x + c
                     2
Out[15]= c + b x + a x
In[16]:= Roots[quadratic == 0, x]
                     2                           2
         -b + Sqrt[b  - 4 a c]        -b - Sqrt[b  - 4 a c]
Out[16]= x == --------------------- || x == ---------------------
                    2 a                              2 a
```

The function `Factor` applies as well to symbolic expressions with more than one variable. For example (make sure you have cleared y!):

```
In[17]:= Factor[3 x^3 + 3 x^4 - 3 x y - 2 x^2 y - 5 x^3 y - 6 x^4 y
         - y^2 + 5 x y^2 + 4 x^2 y^2 - 2 x^3 y^2 + 2 y^3 + 2 x y^3]
                        2
Out[17]= -((1 + x) (x  - y) (3 x + y) (-1 + 2 y))
```

1.4. SYMBOLS

If you question *Mathematica* about `Factor` you will find that `Factor` is limited to factoring polynomial expressions *over the integers*. This means that neither of the following factorizations would be produced by `Factor`, although you may well want to use them sometime:

$$x^2 - 5 = (x - \sqrt{5})(x + \sqrt{5}), \text{ or } x^2 + 1 = (x - i)(x + i)$$

```
In[18] := Factor[x^2 - 5]
              2
Out[18] = -5 + x
```

Similarily, invoking the function `Simplify` will not automatically give the simplest form possible—*Mathematica* only returns the simplest form it finds. Take for instance $\log(e^x)$. In a calculus course in which it can be assumed x is a real number, $\log(e^x) = x$. But *Mathematica* does not limit itself to real numbers, and if complex numbers are permitted then it is not necessarily true that $\log(e^x) = x$.

```
In[19] := Log[E^x]
              x
Out[19] = Log[E ]
In[20] := Simplify[%]
              x
Out[20] = Log[E ]
```

Simplifying trigonometric expressions poses a similar problem, but *Mathematica* does offer a special Package to deal with this. (See Section 1.8 below.) As an alternative, you can try simplifying trigonometric expressions by putting trigonometric identities in the form of replacements. For example:

```
In[21] := Sqrt[1 - Sin[x]^2] /. 1 - Sin[x]^2 -> Cos[x]^2
Out[21] = Cos[x]
```

Finally, it is important to realize that when defining a function — in any mathematical context, not just in *Mathematica* — the variable, say x, used as the argument of the function is a *dummy* variable. In *Mathematica* this is indicated by the subscript `_`, called a *blank*, on `x`. (Recall that when you define a function $f(x)$ in *Mathematica* you do it by entering `f[x_] := ...` .) The letter name of a blank, or dummy, variable is irrelevant; any variable will do. If, for example, we wish to define a function

$$f(x) = x^3 + 5x - 6,$$

we could do it by

```
In[22] := f[x_] := x^3 + 5 x - 6
```

or you could just as well use

```
In[23] := f[s_] := s^3 + 5 s - 6
```

Either way, we obtain the same pattern:

$$f(anything) = (anything)^3 + 5(anything) - 6.$$

Suppose that you have defined f as in `In[22]`, using `x_`. Now also suppose $z = ax + b$ and $g(x) = zf(x)$. What is $g(2)$? Obviously $g(2) = 12(2a + b)$, you say. Try these calculations in *Mathematica*, however:

```
In[24] := z = a x + b
Out[24] = b + a x
In[25] := g[x_] := z f[x]
In[26] := g[2]
Out[26] = 12 (b + a x)
```

Note that *Mathematica*, rightfully, does not equate the variable `x` used in `In[24]` with the blank `x_` used to define g in `In[25]`. Watch out for such *Mathematical* behaviour — it is a little more rigorous and precise than what you might be accustomed to.

MATHEMATICA FUNCTIONS INTRODUCED IN THIS SECTION
Expand   Factor   Simplify
Roots

Exercises:

1. Factor $x^3 + 4x^2 + x - 6$.

2. Expand $(1 + a)(b + a^2)(b^2 - 1)$.

3. Factor $x + \dfrac{2}{x} + \dfrac{1}{x^2} - \dfrac{2}{x-1}$.

4. (Product Rule) Differentiate the product $f(x)g(x)$ with respect to x.

5. (Quotient Rule) Differentiate the quotient $\dfrac{f(x)}{g(x)}$ with respect to x.

6. (Chain Rule) Differentiate the composition $f(g(x))$ with respect to x.

7. Set $y = 2x(x^2 + 1)$, and find the value of y if x were set equal to the following expressions (you may wish to simplify your answers):

(a) 27

(b) -1

(c) $a + b$

(d) $\sqrt{5}$

(e) $\dfrac{1+s}{1-s}$

(f) What happens if you try differentiating y by using `y'` ?

(g) Differentiate y by using `D[y, x]`.

8. Define a function f such that $f(x) = 2x(x^2 + 1)$.

    (a) What happens if you try differentiating f by using `D[f, x]`? Why?

    (b) Find the derivative of f with respect to x.

    (c) Find $f'(3)$.

9. Can you use y, as defined above, and a single line of input, to find $f'(3)$, without using `f` in your input?

10. Find $f''(3)$.

11. Can you use y, as defined above, and a single line of input, to find $f''(3)$, without using `f` in your input?

12. Let $s(n) = 1^2 + 2^2 + 3^2 + \cdots + n^2$. Prove

$$(1.1) \qquad s(n) = \frac{n(n+1)(2n+1)}{6}$$

by induction on n. Do all your calculations in *Mathematica*. (That is, show (1.1) is true for $n = 1$; and that if (1.1) is true for n then it is also true for $n + 1$.)

1.5 Lists and Tables

Thus far we have seen that *Mathematica* uses round brackets, ( ), to group algebraic expressions and uses square brackets, [ ], to enclose the argument(s) of functions. Your keyboard has one other type of brackets, known as brace brackets, { }. *Mathematica* uses these to enclose objects in lists. For instance

`In[1]:= {Sin[Pi], Sin[3], Log[7], Exp[-2], Sqrt[17]}`

allows *Mathematica* to handle five inputs at once:

Out[1]= {0, Sin[3], Log[7], E^{-2}, Sqrt[17]}

Most functions in *Mathematica* are *listable*; a listable function acting on a list evaluates each object in the list separately. So we can approximate each of the above five expressions at once by

In[2]:= % // N
Out[2]= {0., 0.14112, 1.94591, 0.135335, 4.12311}

A list of data can be used to generate a table of function values. For instance, a table of log(n!) for n from 1 to 10:

In[3]:= list = {1!, 2!, 3!, 4!, 5!, 6!, 7!, 8!, 9!, 10!}
Out[3]= {1, 2, 6, 24, 120, 720, 5040, 40320, 362880, 3628800}
In[4]:= Log[list] // N
Out[4]= {0., 0.693147, 1.79176, 3.17805, 4.78749, 6.57925, 8.52516, 10.6046, 12.8018, 15.1044}

Lists such as these can be more conveniently generated by using *Mathematica*'s Table function and an iterator:

In[5]:= Table[i!, {i, 1, 10}]
Out[5]= {1, 2, 6, 24, 120, 720, 5040, 40320, 362880, 3628800}

There are four possible ways of defining an iterator; each way uses one or more objects enclosed as a list in brace brackets:
{ n } means: iterate n times, without incrementing any variables.
{ i, imax } means: i goes from 1 to $imax$ in steps of 1.
{ i, imin, imax } means: i goes from $imin$ to $imax$ in steps of 1.
{ i, imin, imax, d } means: i goes from $imin$ to $imax$ in steps of d.
Thus the following input will generate a table of log(n!) for n running through the even numbers from 2 to 10:

In[6]:= Table[Log[i!], {i, 2, 10, 2}] // N
Out[6]= {0.693147, 3.17805, 6.57925, 10.6046, 15.1044}

(By the way, if you prefer the data in a list to be presented vertically, instead of horizontally, use the *Mathematica* function TableForm instead of Table.)

Iterators can also be used for finite sums, in conjunction with the function Sum. So for instance,

$$\sum_{i=1}^{10} i^2$$

is entered into *Mathematica* as

1.5. LISTS AND TABLES

```
In[7] := Sum[i^2, {i, 1, 10}]
Out[7]= 385
```

The sum of the first five odd numbers could be entered as:

```
In[8] := Sum[i, {i, 1, 10, 2}]
Out[8]= 25
```

Note that even though $imax = 10$, *Mathematica* only adds up the odd numbers from 1 to 9. Similarily, for an input such as `Sum[i, {i, 3.5 }]`, where step size is 1 (by default), the output would be 6, ie

$$\sum_{i=1}^{3} i = 6,$$

since 3 is the greatest integer less than or equal to 3.5.

The limits of integration in a definite integral are indicated by `{ x, xmin, xmax }`, which is very similar to the syntax for iterators. But in this case there is no specification for step size, since integration is performed in terms of intervals of real numbers, not discrete sums. Thus

$$\int_0^1 x^3 dx$$

is entered into *Mathematica* by

```
In[9] := Integrate[x^3, {x, 0, 1}]
            1
Out[9]= -
            4
```

Limits of integration may be *variables*, but the limits for iterators, as used with `Table` and `Sum` must be *numerical*, if simplification is to occur. Thus

$$\int_a^b x dx$$

can be entered as

```
In[10] := Integrate[x, {x, a, b}]
                 2     2
              -a     b
Out[10]=    ---  +  --
                 2     2
```

and the correct symbolic output appears. But

$$\sum_{i=1}^{n} i$$

is not simplified
: `In[11] := Sum[i, i, 1, n]`
`Out[11]= Sum[i, i, 1, n]`
unless a value of n is indicated, say by `/.` `n -> 5`, which would result in an output of 15.

Mathematically the two lists $\{1, 2\}$ and $\{2, 1\}$ are equal as sets, but *Mathematica* actually considers each list as an *ordered* set. Thus:
`In[12] := {1, 2} == {2, 1}`
`Out[12]= False`
Therefore the ordered pair (x, y) can be represented in *Mathematica* by the list {x, y}; the ordered triple (x, y, z), by the list {x, y, z}; etc. Indeed, in terms of Linear Algebra, *vectors* are represented in *Mathematica* by lists, *matrices* by lists of lists, and so on. But as this is a Calculus course we will not be using these last two interpretations of lists very often.

To extract the *ith* object in a list you must use `[[i]]` :

`In[13] := a = Table[i^2, {i, 1, 10}]`
`Out[13]= {1, 4, 9, 16, 25, 36, 49, 64, 81, 100}`
`In[14] := a[[3]]`
`Out[14]= 9`

And to extract more than one object from a list, say the third and sixth:

`In[15] := a[[{3, 6}]]`
`Out[15]= {9, 36}`

which gives a short list of two objects. If you have a list of lists, say

$$b = \{\{2, 3, 6\}, \{5, -1, 7\}, \{4, 7, -3\}\},$$

then the second entry in the third list of b could be extracted by
`In[16] := b[[3]][[2]]`
`Out[16]= 7`

since `b[[3]]` , the third entry of b, is itself a list. In cases such as this, *Mathematica* allows you to shorten an expression such as `In[16]` by entering `b[[3, 2]]` . Finally, functions that you define yourself are also listable.

1.6. SOLVING EQUATIONS

MATHEMATICA FUNCTIONS INTRODUCED IN THIS SECTION
Table   Sum

Exercises:

1. Define a to be the list of $\sin 1, \sin 2, \ldots, \sin 10$. Do not approximate your values.
2. Calculate decimal approximations for each number in a.
3. Generate a list of the first 10 square numbers by using the `Table` function.
4. Square a list of the first 10 numbers; compare with result of Exercise 3.
5. Make a table of the first 100 prime numbers.
6. Calculate $\sum_{i=1}^{100} i^2$.
7. Sum the arithmetic series $3 + 7 + 11 + \cdots + 67$.
8. Calculate $\int_0^\pi \sin x\, dx$.
9. Evaluate $\int_{s+t}^{st} \sin x\, dx$.
10. Define a function f such that $f(x) = \sqrt{1 + x^3}$.
11. For f as in Exercise 10 and a as in Exercise 1, compute $f(a)$.
12. Approximate the 7*th* entry of the list in Exercise 11.
13. Global human population is approximated by the equation
$$P(t) = 5e^{0.017t},$$
where $P(t)$ is the population in *billions*, t years since 1986. Construct a table showing the population figures, every 10 years, from 1936 to 2036.

1.6 Solving Equations

To solve an equation with *Mathematica* the basic command is
$$\texttt{Solve[lhs == rhs, x]}.$$
This will solve the equation $lhs = rhs$ for x, and present the solution(s) as a list of replacement values for x. *Mathematica* will, of course, try to give exact solutions — but this will not always be possible. However, by using `N`, it will usually be possible to obtain approximate solutions. Examples follow:

```
In[1]:= Solve[x^2 + x - 2 == 0, x]
Out[1]= {{x -> -2}, {x -> 1}}
```

Observe that this solution set is presented as a list of substitutions by *Mathematica* and so its constituent parts may be used in future calculations by extracting the desired part(s). For instance, what is the value of $x^3 + x - 6$ for the second solution to the above equation?

```
In[2]:= x^3 + x - 6 /. %[[2]]
Out[2]= -4
```

It is possible to solve equations with literal coefficients:

```
In[3]:= Solve[a x + b == 0, x]
                    b
Out[3]= {{x -> -(-)}}
                    a
```

Here are two examples with which you should be quite familiar as they can both be solved using the standard quadratic formula:

```
In[4]:= Solve[x^2 - x - 1 == 0, x]
              1 + Sqrt[5]          1 - Sqrt[5]
Out[4]= {{x -> -----------}, {x -> -----------}}
                   2                    2
In[5]:= % // N
Out[5]= {{x -> 1.61803}, {x -> -0.618034}}
In[6]:= Solve[x^2 + 1 == 0, x]
Out[6]= {{x -> I}, {x -> -I}}
```

Formulas do exist for solutions to cubic, quartic and some other higher degree equations, but they are usually so cumbersome that they are hardly ever used in practice. With a computer, though, these convoluted formulas can be used, and the exact — though sometimes exceedingly messy — solutions can be found:

```
In[7]:= Solve[x^5 + x^3 - 4 x + 2 == 0, x]
Out[7]= ...  pages of output scroll by  ...
```

You can scroll through the formulas for the five exact solutions, if you are interested; to approximate them, simply use % // N :

```
In[8]:= % // N
Out[8]= {{x -> 1.}, {x -> 0.556391}, {x -> -1.37426},

   >    {x -> -0.0910667 + 1.61474 I}, {x -> -0.0910667 - 1.61474 I}}
```

1.6. SOLVING EQUATIONS

Such approximations can be found even if *Mathematica* cannot find explicit algebraic solutions:

```
In[9] := Solve[x^5 - x^3 + x^2 == 3, x]
Out[9]= {ToRules[Roots[x   - x  + x  == 3, x]]}
                      2    3    5
In[10] := % // N
Out[10]= {{x -> -1.16131 - 0.471424 I}, {x -> -1.16131 + 0.471424 I},
{x->0.520283 -1.10405 I},{x->0.520283+1.10405 I},{x->1.28205}}
```

In some cases, even though there is a solution, *Mathematica* will not present any solutions at all. Try:

```
In[11] := Solve[Exp[-x] == x, x]
Solve::ifun: Warning: inverse functions are being used by Solve,
so some solutions may not be found.
Solve::tdep: The equations appear to involve transcendental functions
of the variables in an essentially nonalgebraic way.
              -x
Out[11]= Solve[E   == x, x]
In[12] := % // N
Solve::ifun: Warning: inverse functions are being used by Solve, so
some solutions may not be found.
Solve::tdep: The equations appear to involve transcendental functions
of the variables in an essentially nonalgebraic way.
                    -x
Out[12]= Solve[2.71828   == x, x]
```

You get some informative messages, but no solutions. In addition, entering % // N will get you more messages — but no approximate solutions. However, as we shall see in Chapter 4, there are other ways to find approximate solutions to any equation, both in theory, and with *Mathematica*.

To solve a system of equations, use

$$\text{Solve}[\textit{list of equations, list of variables}].$$

The solution will be presented as a list of lists. For example, consider the system

$$\begin{aligned} x + y + z &= 6 \\ 2x - y + z &= 5 \\ -3x + y + 2z &= 6 \end{aligned}$$

which has solution $(x, y, z) = (1, 1, 4)$. In *Mathematica*:

```
In[13]:= Solve[{x + y + z == 6, 2 x - y + z == 5, -3 x + y + 2 z == 6},
                {x, y, z}]
Out[13]= {{x -> 1, y -> 1, z -> 4}}
```

In addition to `Solve`, *Mathematica* offers the function `Reduce`, which also acts on systems of one or more equations, but produces slightly different output: namely a complete list of logical cases and simplified equations equivalent to the original system. Consider the simple equation, $ax + b = 0$ which was solved in `Out[3]` above: $x = -\dfrac{b}{a}$. This solution is meaningless if $a = 0$, but `Solve` does not take such special cases into account. However, `Reduce` does:

```
In[14]:= Reduce[a x + b == 0, x]
                        b
Out[14]= a != 0 && x == -(-) || a == 0 && b == 0
                        a
```

Here the `!=` means $\neq$, and the `&&` and `||` are logical symbols for *and* and *or*, respectively.

> **MATHEMATICA FUNCTIONS INTRODUCED IN THIS SECTION**
> Solve   Reduce

Exercises:

1. Solve the system
$$\begin{aligned} x + 3y - 4z &= 7 \\ 5x - y - 5z &= 12 \\ x + y + z &= 14 \end{aligned}$$

2. Solve the general system of two linear equations in two unknowns:
$$\begin{aligned} ax + by &= e \\ cx + dy &= f \end{aligned}$$

3. Reduce the general system of two linear equations in two unknowns — scroll through the output.

4. Solve the equation $x^3 + 4x^2 = 6$ for x.

5. Approximate the solutions from Exercise 4.

6. Solve the equation $3.1^x = 14.2$ for x. (*Mathematica*'s answer is somewhat of a mystery!)

1.7. GRAPHING

7. Approximate the answer to Exercise 6.

8. Solve the equation $x^3 + 45x^2 - 574x - 4007 = 0$ for x. Scroll through the output.

9. Approximate the three solutions to Exercise 8. Note the small ($\approx 10^{-15}$) imaginary component in two of the solutions, even though all three solutions are actually real numbers.

10. Solve for the intersection points of the ellipse with equation $4x^2 + 9y^2 = 25$ and the hyperbola with equation $x^2 - y^2 = -1$.

11. Try solving the equation $\sin x = 1$ for x. Note that *Mathematica* gives only one of infinitely many solutions.

12. Solve the system
$$\begin{aligned} x + y - 2z &= 4 \\ 2x - y + z &= 12 \end{aligned}$$
for x, y, and z. Note that since there are infinitely many solutions, *Mathematica* gives parametric solutions, writing x and y in terms of z.

13. By using **Sum** and **Solve**, try solving the equation
$$\sum_{i=1}^{n} i = 15.$$
The correct answer is $n = 5$, but *Mathematica* produces no answer at all.

14. Define a polynomial function p such that
$$p(x) = ax^2 + bx + c.$$

   (a) Find a, b, and c, such that
$$p(0) = 3, p(1) = -2, p(3) = -6.$$

   (b) Calculate the values of $p(2)$ and $p(7)$.

15. Try solving the equation
$$|x| = 4$$
which has the two obvious solutions $x = 4$ and $x = -4$. Note that *Mathematica* produces no solutions at all.

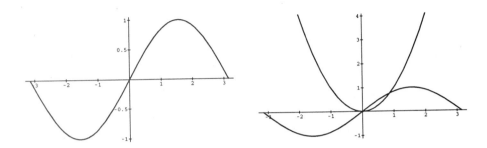

Figure 1.1: Basic plotting with *Mathematica*: $\sin x$ on the left; $\sin x$ and x^2 on the right.

1.7 Graphing

To produce the graph of the function $y = f(x)$ on part of its domain, say the interval from $x = a$ to $x = b$, the basic *Mathematica* function is Plot[f[x], {x, a, b}], or Plot[y, {x, a, b}]. For example, consider the graph of $y = \sin x$, on the interval $[-\pi, \pi]$.

```
In[1]:= Plot[Sin[x], {x, - Pi, Pi}]
Out[1]= -Graphics-
```

Your screen should show something like Figure 1.1, left part. To produce simultaneously the graphs of two or more functions, plot a list of functions:

```
In[2]:= Plot[{x^2, Sin[x]}, {x, - Pi, Pi}];
```

Your output should look something like Figure 1.1, right part. (Note also that ending the input line with ; prevents the display of Out[2] = -Graphics-.) Consider, however, the following:

```
In[3]:= list = { Exp[x], Exp[-x], Sin[x], x^2}
              x    -x            2
Out[3]= {E  , E  , Sin[x], x }
In[4]:= Plot[list, {x, - Pi, Pi}];
Plot::notnum: list does not evaluate to a real number at x=-3.14159.
Plot::notnum: list does not evaluate to a real number at x=-2.87979.
```

1.7. GRAPHING

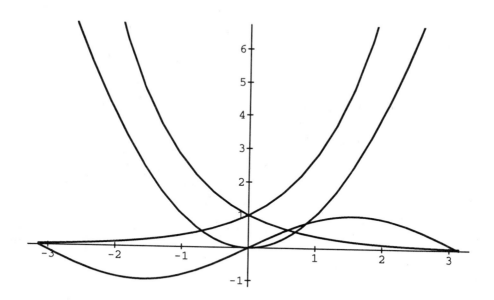

Figure 1.2: Four graphs plotted at once: $\sin x$, x^2, e^x, and e^{-x}. Which is which?

```
Plot::notnum: list does not evaluate to a real number at x=-2.61799.
General::stop: Further output of Plot::notnum
     will be suppressed during this calculation.
```

The problem here is that *Mathematica* is trying to generate a table of values from the symbol `list` alone, without evaluating any of the actual functions in the list. To remedy this, use the function `Release`, which tells *Mathematica* first to make the list of functions, and then to evaluate them for particular values of x. Your output should look something like Figure 1.2.

```
In[5]:= Plot[Release[list], {x, - Pi, Pi}];
```

Note that in using `Plot` you must specify the interval of values to be considered. If you know what the graph looks like, then you can probably pick a suitable interval to get a suitable picture of the graph. However, if you don't know beforehand what the graph looks like, then it may not be easy to find a suitable interval over which to sketch the graph. Consider the graph of $y = x^3 + 45x^2 - 574x - 4007$. If you plot it for x between -4 and 4 the graph seems almost like a straight line. See Figure 1.3, top part.

```
In[6]:= Plot[x^3 + 45 x^2 - 574 x - 4007, {x, -4, 4}];
```

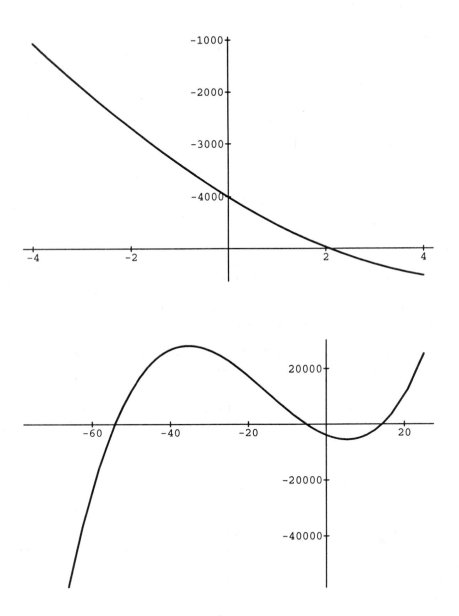

Figure 1.3: The graph of $y = x^3 + 45x^2 - 574x - 4007$ on two different domains: on $[-4, 4]$ (top); and on $[-75, 25]$ (bottom).

1.7. GRAPHING

If you recall your high school calculus, then you will realize that we should pick an interval which includes the critical values of y. These can be found by solving the equation $y' = 0$. The following sequence of steps shows how easy this kind of investigative graphics is in *Mathematica*:

```
In[7]:= D[x^3 + 45 x^2 - 574 x - 4007, x]
                               2
Out[7]= -574 + 90 x + 3 x
In[8]:= Solve[% == 0, x] // N
Out[8]= {{x -> 5.40425}, {x -> -35.4042}}
```

So we could now take our x-values from -50 to 10, say:

```
In[9]:= Plot[x^3 + 45 x^2 - 574 x - 4007, {x, -50, 10}];
```

From -75 to 25 would probably be even better. See Figure 1.3, bottom part.

```
In[10]:= Plot[x^3 + 45 x^2 - 574 x - 4007, {x, -75, 25}];
```

The interval over which you specify the graph to be plotted can also severely effect the scale of the graph. Consider the graph of $y = x^3 - 3x - 4$. See Figure 1.4.

```
In[11]:= Plot[x^3 - 3x -4, {x, -5, 5}];
In[12]:= Plot[x^3 - 3x -4, {x, -1000, 1000}];
```

In one case the interesting behaviour around the origin is highlighted (Figure 1.4, top part); in the other case, it is completely obliterated. (Figure 1.4, bottom part.)

Mathematica will produce a graph if the function is defined everywhere except at some isolated points. In both of the following examples the function is defined everywhere except at $x = 0$.

```
In[13]:= Plot[1/x, {x, -2, 2}];
In[14]:= Plot[Sin[1/x], {x, -2, 2}];
```

If you wish to plot the graph of a relation (that is, a set of ordered pairs) which is not a function[1] then you must give parametric equations for x and y and use `ParametricPlot`. For example, the circle of radius 3 with centre at the origin has parametric equations

$$\begin{cases} x = 3\cos t \\ y = 3\sin t \end{cases}$$

where the parameter t represents the angle in radians. This can be plotted as:

---

[1] This is a good time to remind yourself about the difference between a relation and a function; and what extra conditions are needed for a function to be injective (one-to-one), surjective (onto), or both (bijective).

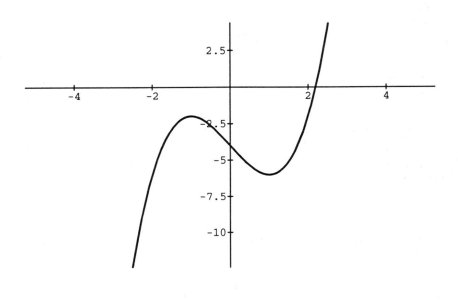

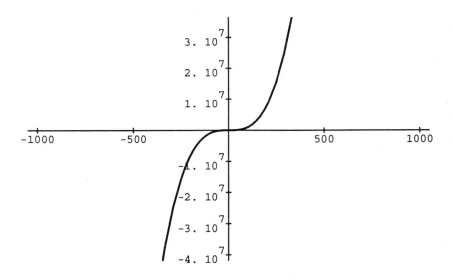

Figure 1.4: The graph of $y = x^3 - 3x - 4$ to two different scales.

1.7. GRAPHING

```
In[15]:= ParametricPlot[{3 Cos[t], 3 Sin[t]}, {t, 0, 2 Pi}];
```

However, the result doesn't *look* like a circle! See Figure 1.5, top part. This is because *Mathematica* has a default setting of height–to–width ratio (called the `AspectRatio`) of about 0.618.[2] You can alter this ratio by using an *option* of `ParametricPlot`, namely `AspectRatio -> Automatic`, which sets the height–to–width ratio at 1. Then your output should look more like a circle. See Figure 1.5.

```
In[16]:= ParametricPlot[{3 Cos[t], 3 Sin[t]}, {t, 0, 2 Pi},
         AspectRatio -> Automatic];
```

There are many additional options that can be used with `Plot` and `ParametricPlot`; some of them will be explored in the exercises. There is another variation of `Plot`, called `ListPlot`, which allows you to plot isolated points:
`ListPlot[{`$y_1, y_2, y_3, \ldots$`}]` will plot the points

$$(1, y_1), (2, y_2), (3, y_3), \ldots,$$

while `ListPlot[{{`x_1, y_1`}, {`x_2, y_2`}, {`x_3, y_3`}, ...}]` will plot the points

$$(x_1, y_1), (x_2, y_2), (x_3, y_3), \ldots.$$

If you wish to construct a graph passing through isolated points, you can use the `PlotJoined -> True` option of `ListPlot`. However, this will only join consecutive dots with straight lines. See Figure 1.6, top part.

```
In[17]:= data = {{1, 1.01}, {2, 3.97}, {4, 16.3}, {6, 34.7}}
Out[17]= {{1, 1.01}, {2, 3.97}, {4, 16.3}, {6, 34.7}}
In[18]:= ListPlot[%];
In[19]:= ListPlot[data, PlotJoined -> True];
```

In case you have numerical data, you can obtain an approximate functional expression by using the *Mathematica* function `Fit`. This fits any family of functions to your data, and will result in a smoother fit than simply joining consecutive data points by straight lines. For example, consider finding a quadratic function — a combination of 1, x and x^2 — to fit the above data:

```
In[20]:= Fit[data, {1, x, x^2}, x]
                                              2
Out[20]= -0.833869 + 0.847098 x + 0.847425 x
In[21]:= Plot[%, {x, 1, 6}];
```

---

[2]Can you guess why?

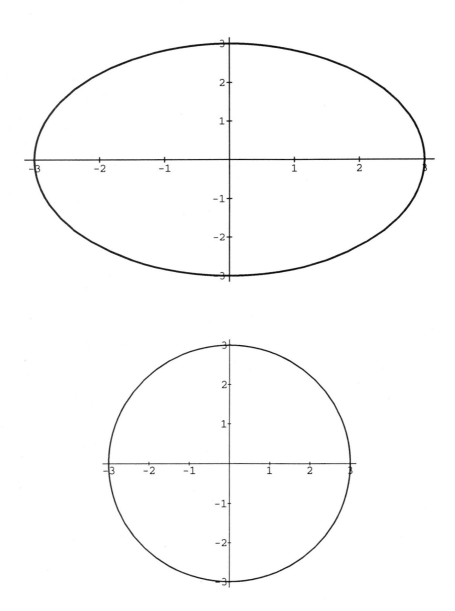

Figure 1.5: The graph of the circle $x^2 + y^2 = 9$ with two different aspect ratios; only with the option `AspectRatio -> Automatic` does the circle look like a circle.

1.7. GRAPHING

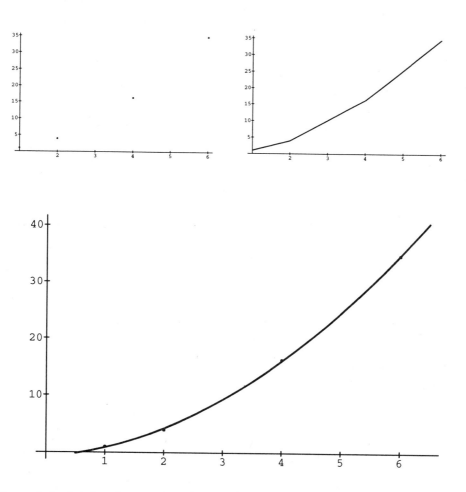

Figure 1.6: Joining data points (top left) with straight line segments by using the option `PlotJoined -> True` (top right); fitting a curve to data points by using the function `Fit` (bottom).

As you can see this seems like a much smoother fit to the given data points. In fact, you could compare the quadratic of `Out[20]` and the given data points, by displaying the graphs of `Out[21]` and `Out[18]` together. (This can be done with `Show` — see below.) Note that the graph does not actually pass through any of the given points. See Figure 1.6, bottom part.

```
In[22]:= Show[%, Out[18]];
```

When using `Plot`, or any of its variations, *Mathematica* will accept some pretty complicated expressions as arguments. Consider for instance the function $f(x)$ defined in terms of the following two cases:

$$f(x) = \begin{cases} x^2 + 1, & \text{if } x \geq 0 \\ x, & \text{if } x < 0 \end{cases}$$

In *Mathematica* f can be defined via a logical `If` statement — and its graph is plotted with no apparent difficulty.

```
In[23]:= f[x_] := If[x < 0, x, x^2 + 1]
In[24]:= Plot[f[x], {x, -2, 2}];
```

Or consider an example in which the function to be plotted is defined in terms of a previous output line:

```
In[25]:= {x^2, x^3, x^4}
                2   3   4
Out[25]= {x , x , x }
In[26]:= Plot[Out[25][[1]], {x, -2, 2}];
```

Again, *Mathematica* has no difficulty in plotting the graph. However, should you ever be working with an expression f which is so complicated that *Mathematica* is not able to plot its graph, you should try `Plot[Release[f], ...]`. If this fails to produce results, re-examine the definition of your function!

Once you have produced a graphics object you can review it any time you wish simply by using `Show`:

```
In[27]:= Show[Out[10]];
```

The great advantage of using `Show` is that you do not have to recreate the whole definition of the plot. In addition, `Show` allows you to change the options of a plot one by one to see their effect.

```
In[28]:= Show[%, PlotLabel -> "graph of a cubic"];
```

1.7. GRAPHING

`Show` also allows you to combine previous graphics objects into one new object:

```
In[29]:= Show[Out[26], Out[16]];
In[30]:= Show[Out[26], Out[16], AspectRatio -> Automatic];
```

Poducing graphic images is one of the strong points of *Mathematica* and can be invaluable in investigating the shape of functions, or in finding approximate solutions to complicated equations. Beware, however, that very different functions can have locally similar graphs.

Graphs that you have created are lost once you end your session, unless you save them on the default directory by using `Display`. This function will save a *PostScript* form of your plot, in a file of your choice, for future viewing. The syntax is `Display["filename", plot]`. If you are working in a DOS environment, you can then view the graph so saved by entering the command `display filename` from the DOS prompt.) *Mathematica* does not produce *PostScript* in the encapsulated form that is needed for export to document preparation systems like LaTeX or laserprinters external to *Mathematica*, but it is easily so converted. For example, in UNIX the program `psfix` puts the necessary header information on the file saved by `Display`; in DOS the external program `PRINTPS` is used for printing or for preparing in encapsulated form the graphics objects for export. Finally, even though this is a first-year calculus course of functions in one variable, we should mention that *Mathematica* can produce three-dimensional coloured plots as well, via the function `Plot3D`:

```
In[31]:= Plot3D[Sin[3 x y], {x, 0, Pi}, {y, 0, Pi}, PlotPoints -> 40,
            Lighting -> True]
Out[31]= -SurfaceGraphics-
```

MATHEMATICA FUNCTIONS INTRODUCED IN THIS SECTION		
Plot	Release	ParametricPlot
ListPlot	Show	Fit
Display	Plot3D	

Exercises:

1. Find a suitable interval on which to plot the given function:

    (a) $y = x^6 - 47x^4 + 304x^2 - 5007$

    (b) $y = \left(\log \sqrt{\dfrac{14 + 7x}{3 + 10x^2}}\right)^2$

    (c) $y = x^2 e^x$

(d) $y = x - \sin x$

2. Plot graphs of
$$x^2 - 3x + 2, \quad x^2 + x + 3, \quad \frac{x^2}{2} + 3x - 5$$
simultaneously on intervals ranging from $[-4, 4]$ to $[-400, 400]$. What do you notice?

3. Plot graphs of
$$\frac{2x^3 + 5x - 2}{x^2 + 3x + 4}, \quad 2x$$
simultaneously on intervals ranging from $[-5, 5]$ to $[-500, 500]$. What do you notice?

4. Plot graphs of the three functions
$$f(x) = e^{-x/4} \sin \pi x, \quad g(x) = e^{-x/4}, \quad h(x) = -e^{-x/4}$$
on the same set of axes; what do you notice?

5. Plot graphs of $y = x^2$ and $y = \sin x$ on the same set of axes to determine how many real solutions there are to the equation
$$x^2 = \sin x.$$

6. Investigate the graph of $x^2 \sin \frac{1}{x}$ on different intervals ranging from $[-10, 10]$ to $[-0.01, 0.01]$.

7. Plot the graph of $f(x) = x^3 - 3x + 2$ on the interval $[-3, 4]$; enter it as `graph = Plot[x^3 - 3 x + 2, {x, -3, 4}]`.

    (a) View the previous graph once again by entering `Show[graph]`.

    (b) Investigate some options of `Plot` by entering the following:

    i. `Show[graph, AspectRatio -> 2]`
    ii. `Show[graph, AspectRatio -> 1/2]`
    iii. `Show[graph, PlotLabel -> "y = x^3 - 3x + 2"]`
    iv. `Show[graph, AxesLabel -> {"x value", "x^3 - 3x + 2"}]`
    v. `Show[graph, Framed -> True]`

8. What are the parametric equations of an ellipse, centred at $(x, y) = (1, -3)$, with semi–major axis 4, and semi–minor axis 3?

    (a) Use `ParametricPlot` to sketch the above ellipse; include the option `AspectRatio -> Automatic` .

1.7. GRAPHING

(b) Enter Show[graph, %] ; what do you notice? Now reverse the order, so that *graph* is the second argument of Show; what do you notice?

9. Use ListPlot to plot the first 5 terms of the sequence $a_n = 1/n^2$.

10. Use ListPlot to plot the points on the graph of $y = x^2$ for which x is an even number between 0 and 10, inclusive.

11. Plot $y = x^2$ on the interval $[-2, 10]$.

12. Use Show to combine your plots from the two previous exercises; this is one way to highlight certain points on a graph.

13. Construct a graph of $y = x^3 - 3x + 2$ on which all critical points and inflection points are highlighted.

14. Define a function f such that

$$f(x) = \begin{cases} -x+2 & \text{if } x > 1 \\ x-2 & \text{if } x \leq 1 \end{cases}$$

(a) Plot $f(x)$ over the interval $[-2, 3]$. What do you notice about the graph that is incorrect? What does $f(1)$ actually equal?

(b) Highlight the point $(1, f(1))$ on the previous graph.

15. Define a function $s(n)$ such that

$$s(n) = \sum_{i=1}^{n} i.$$

(a) Plot the graph of s on the interval $[1, 20]$. How can you explain the step–nature of this graph?

(b) The remaining exercises of this question will determine *graphically* a formula for $s(n)$.

  i. By plotting four graphs at once on the interval $[1, 20]$ and looking at the ouput, decide which power of x the graph of $s(x)$ most resembles, x, x^2, or x^3?

  ii. Plot the graph of $\frac{s(x)}{x^2}$ on the interval $[1, 100]$. To which value of y does the graph appear to be approaching as x gets larger and larger? Conclude $s(x) \approx x^2/2$.

  iii. Now compare the plot of $s(x) - x^2/2$ with plots of x and x^2, for x going from 1 to 40.

iv. Now plot $\frac{(s(x) - \frac{x^2}{2})}{x}$ for x going from 1 to 40. Note that the values fluctuate periodically between -0.5 and 0.5. Conclude $s(n) \simeq \frac{n^2}{2} + \frac{n}{2}$.

(c) Test this approximation by plotting $s(x)$ and $(x^2+x)/2$ on the same axes, say for x going from 1 to 40. How would you describe the fit? (Of course, having arrived at a possible formula for $s(n)$, you could now prove it by induction.)

16. Find a cubic polynomial fit to $\log n!$ over the domain $n = 1, 2, \ldots, 20$. Plot the points and the polynomial.

1.8 Packages

Mathematica offers many pre–loaded packages. These are `*.m` files stored in subdirectories of the directory `/math/Packages`. These packages offer many additional functions, but to access them in any *Mathematica* session you must load the package first. The best way is to use `<<` if you give the file's full pathname — probably something like `/usr/lpp/math/Packages/`*filename*. If the package is in the default directory, then it is also possible simply to use `<<` *filename*.[3] Another way of doing this is to use the *Mathematica* function `Needs`. Query *Mathematica* on your terminal about these alternatives using `?Needs` and `?<<`.

By way of example, the package `Trigonometry.m` in the subdirectory `Algebra`, contains a function `TrigReduce` which can simplify many more trigonometric expressions than `Simplify`.

```
In[1]:=  << Algebra/Trigonometry.m
In[2]:= ?TrigReduce
TrigReduce[expr] writes trigonometric functions of  multiple angles
as sums of products of trigonometric functions of  that angle.
In[2]:= TrigReduce[Sin[a + b]]
Out[2]= Cos[b] Sin[a] + Cos[a] Sin[b]
In[3]:= TrigReduce[Sin[x]/Cos[x]]
Out[3]= Tan[x]
In[4]:= TrigReduce[Sqrt[1 - Sin[x]^2]]
Out[4]= Cos[x]
```

There are also animation packages on some platforms. In DOS `Animation.m` in the

---

[3] We shall omit the full pathname. If you call in a package which includes usage statements, *Mathematica* will print the last such statement upon succesfully loading the package.

1.8. PACKAGES

subdirectory `Graphics`; it offers functions which allow you to make mathematical movies — a way to show one graph after another in quick succession.

```
In[5]:= << Graphics/Animation.m  (* full pathname
```

```
In[6]:= ?Animate
Animate[{g,h,...}] will animate a list of graphics objects.
```

Once you have created a list of graphics objects, you can use the function `Animate` to "show" the animation sequence again, similar to the way `Show` allows you to review a single previous plot. Animation is also available on NeXT workstations, but we have to wait for *Mathematica* Version 2 for XWindows on UNIX to have animation.

If you have defined some functions which you wish to save for future reference, then you can save them in a package – a *.m file of your own naming – which you can load in your next session via `Needs` or `<<`. Suppose, for instance you are preparing a package on high shool mathematics. You might define the following function, `quadratic`:

```
In[7]:= quadratic[{a_, b_, c_}, x_] := a x^2 + b x + c
In[8]:= quadratic[{22, 35, -12}, s]
                 2
Out[8]= -12 + 35 s + 22 s
```

Having decided it is useful, you may save it in a file `highschool.m` as follows:

```
In[10]:= Save["highschool.m", quadratic]
```

Next, suppose you define another function `ellipse` to plot an arbitrary ellipse:

```
In[11]:= ellipse[{a_, b_}] := ParametricPlot[{a Cos[x], b Sin[x]},
                  {x, 0, 2 Pi}, AspectRatio -> Automatic]
In[12]:= ellipse[{3, 4}];
```

You can now append it to `highschool.m` as follows:

```
In[13]:= Save["highschool.m", ellipse]
```

At this stage, if you enter !!highschool.m, you will see listed on your screen the contents of the file *highschool.m*: the two functions you defined plus their definitions:

```
quadratic/: quadratic[{a_, b_, c_}, x_] := a*x^2 + b*x + c

ellipse/: ellipse[{a_, b_}] :=
    ParametricPlot[{a*Cos[x], b*Sin[x]}, {x, 0, 2*Pi},
                         AspectRatio -> Automatic]
```

You can edit a package like *highschool.m* after exiting *Mathematica*. Or, if you need to make changes during a session, use the shell escape character, ! inside *Mathematica* to prefix your favourite editor; for example:

In[14]:= !vi highschool.m

will invoke the UNIX vi editor and allow you to edit your file. When you quit editing, with :wq to save your changes, you will return to *Mathematica* exactly where you left it. In this book, we have prepared a package for each Chapter, 2 through 12. In each of those chapters you will have to load the package so that the prepared examples and exercises will work out properly.

Finally, feel free to make and save your own packages, especially if you find yourself using a certain routine over and over – define it in terms of functions and save it for future reference.

MATHEMATICA FUNCTIONS INTRODUCED IN THIS SECTION		
Needs	Save	TrigReduce
<<	!!	

Exercises:

1. Call in the package Trigonometry.m and use its function TrigReduce on the following trigonometric expressions.

    (a) $\tan(A+B)$

    (b) $\cos(3A)$

    (c) $\sin(5A)$

    (d) $\sin^4 A$

    (e) $\sin^4 x - \cos^4 x$

    (f) $\dfrac{\cos x \tan^2 x}{\sec x + 1}$

    (g) $\sin(2t)\cos t - \cos(2t)\sin t$

2. Call in the animation package Animation.m. Create an animation sequence of graphs showing the plot of
$$y = kx^2 - e^x$$
for values of k ranging from $k=1$ to 5. Plot each graph on the interval $[-1, 5]$. How many real solutions are there to the equation
$$kx^2 = e^x$$
for each of the values k from 1 to 5; between which consecutive integers does each solution lie?

1.8. PACKAGES

3. Create a file *hook.m* by using the input `!vi hook.m`, and put in it the *Mathematica* function of h as in Exercise 6 of Section 1.3:

$$h(x) = \begin{cases} x, & \text{if } x < 0 \\ x^2, & \text{otherwise} \end{cases}$$

saving it with the command `:wq`. Type it on screen with the command `!!hook.m` and then enter it into *Mathematica* with the input line `<< hook.m`. Find $h(-3)$ and $h(3)$. Note that entering h as a function in this way actually overwrites any previous function called h.

Chapter 2

Beginning Calculus

> *Quantities, and the ratios of quantities, which in any finite time converge continually to equality, and before the end of that time approach nearer to each other than by any given difference, become ultimately equal.* —I.Newton, Philosophiae Naturalis Principia Mathematica 1687, Lemma I.

2.1 Limits That *Mathematica* Can Do

To evaluate limits with *Mathematica* there are many different approaches you can take.

1. Generate tables of numerical data showing values of $f(x)$ as x gets closer and closer to a.

2. Use *Mathematica's* symbolic capabilities to do the usual agebraic manipulation — factoring, simplifying, rationalizing, etc — that you would normally use when evaluating limits with paper and pencil.

3. Plot the function in the neighbourhood of $x = a$ to see what is happening.

4. Use *Mathematica's* `Limit` function.

You should be familiar with all four of these strategies since *Mathematica's* `Limit` function *cannot evaluate all limits that exist*, and *sometimes simply gives the wrong answer!*

Example 1. Consider the limit
$$\lim_{x \to 1} \frac{x^3 - 1}{x^2 - 1}.$$
First we shall ask *Mathematica* to evaluate $f(1)$, then we will illustrate the above four strategies:

```
In[1]:= f[x_] := (x^3 - 1)/(x^2 - 1)
In[2]:= f[1]
                        1
Power::infy: Infinite expression - encountered.
                        0
Infinity::indt: Indeterminate expression 0 ComplexInfinity encountered.
Out[2]= Indeterminate
```

Note that *Mathematica* correctly gives no value for $f(1)$. Let us now calculate a table of values of $f(x)$ as $x \to 1$, making sure to pick some values on each side of 1:

```
In[3]:= f[{1.1, 0.9, 1.05, 0.95, 1.01, 0.99}]
Out[3]= {1.57619, 1.42632, 1.5378, 1.46282, 1.50751, 1.49251}
```

The values of $f(x)$ certainly seem to be approaching a limit of about 1.5. We can show the limit *is* 1.5 in the usual manner by factoring f, dividing out the common factor $x - 1$, and then evaluating the simplified expression at $x = 1$:

```
In[4]:= Factor[f[x]]
                 2
         1 + x + x
Out[4]= ----------
           1 + x
In[5]:= % /. x -> 1
         3
Out[5]= -
         2
```

Or we could use `Limit` and find the limit as follows:

```
In[6]:= Limit[f[x], x -> 1]
         3
Out[6]= -
         2
```

It is interesting to note that in plotting f, *Mathematica* will point out the fact that $f(1)$ is not defined but will create a graph which appears to pass right through the point $(1, 1.5)$, *as if the domain of f included $x = 1$.* See Figure 2.1.

```
In[7]:= Plot[f[x], {x, -1, 3}];
                        1
Power::infy: Infinite expression -- encountered.
                        0.
Plot::notnum: f[x] does not evaluate to a real number at x=-1..
```

2.1. LIMITS THAT MATHEMATICA CAN DO

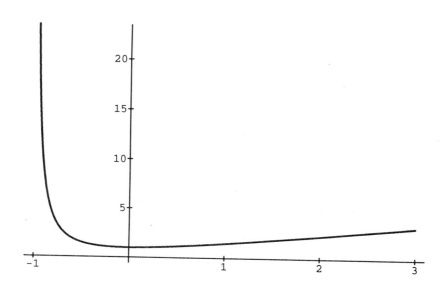

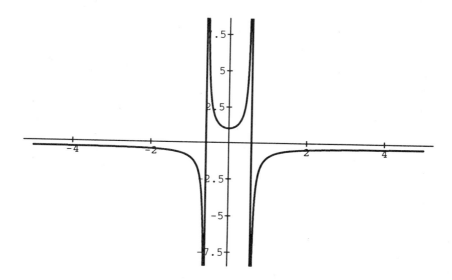

Figure 2.1: Top: *Mathematica* plots $f(x) = \dfrac{x^3 - 1}{x^2 - 1}$ as if $f(1) = 1.5$. Bottom: *Mathematica* plots vertical lines in the graph of $g(x) = \dfrac{x^2 + 1}{1 - 3x^2}$ even though g is not defined at $x = \pm 1/\sqrt{3}$.

Example 2. As a second example, consider
$$\lim_{x \to \infty} \frac{x^2 + 1}{1 - 3x^2}.$$
We can investigate limits at infinity by doing much the same as what we did in Example 1 — we can generate data for x getting larger, we can use `Limit`, and we can plot graphs:

```
In[8]:= g[x_] := (x^2 + 1)/(1 - 3 x^2)
In[9]:= g[{10.0, 100.0, 1000.0}]
Out[9]= {-0.337793, -0.333378, -0.333334}
In[10]:= Limit[g[x], x -> Infinity]
              1
Out[10]= -(-)
              3
```

Note that `Infinity` is used to represent ∞. By using `-Infinity` we can evaluate limits as $x \to -\infty$:

```
In[11]:= Limit[g[x], x -> - Infinity]
              1
Out[11]= -(-)
              3
```

It is also of interest to check what *Mathematica* does for
$$\lim_{x \to 1/\sqrt{3}} g(x),$$
since in this case no limit exists.

```
In[12]:= Limit[g[x], x -> 1/Sqrt[3]]
Out[12]= ComplexInfinity
In[13]:= Limit[g[x], x -> -1/Sqrt[3]]
Out[13]= ComplexInfinity
```

It is not clear what `ComplexInfinity` means, but at least no limit is given. Finally we could plot the graph of g. See Figure 2.1.

```
In[14]:= Plot[g[x], {x, -5, 5}];
```

Note that the graph does have horizontal and vertical asymptotes, as required. However, the fact that *Mathematica* includes vertical lines in the graph of g at $x = \pm 1/\sqrt{3}$ is actually its attempt to join together the different pieces of the graph — it is *not* putting in vertical asymptotes for our convenience.

Example 3. `Limit` does allow you to evaluate many limits that you would otherwise not be able to do, unless you used L'Hôpital's rule. Try the following three limits:

2.1. LIMITS THAT MATHEMATICA CAN DO

```
In[15]:= Limit[(1 + 1/n)^n, n -> Infinity]
Out[15]= E
In[16]:= Limit[x^x, x -> 0]
Out[16]= 1
In[17]:= Limit[x Log[x], x -> 0]
Out[17]= 0
```

```
┌─────────────────────────────┐
│   MATHEMATICA FUNCTIONS     │
│ INTRODUCED IN THIS SECTION  │
├─────────────────────────────┤
│   Limit    Infinity         │
└─────────────────────────────┘
```

Exercises:

1. Use `Limit` to evaluate the following limits. In which cases does *Mathematica* not return the correct answer? What is the correct answer in these cases?

    (a) $\lim_{h \to 0} \dfrac{\sqrt{1-h} - \sqrt{1}}{h}$

    (b) $\lim_{x \to \infty} \dfrac{\sin x}{x}$

    (c) $\lim_{x \to 0} \dfrac{\sin ax}{x}$

    (d) $\lim_{x \to 1} \dfrac{1-x}{1-\sqrt{x}}$

    (e) $\lim_{x \to \infty} x - \sqrt{x^2 + 2x}$

    (f) $\lim_{x \to \infty} x + \sqrt{x^2 + 2x}$

    (g) $\lim_{x \to -\infty} x + \sqrt{x^2 + 2x}$

    (h) $\lim_{x \to \infty} \dfrac{\sqrt{1+x^2}}{x}$

2. Explain why it is actually necessary in the last two limits of Example 3 to assume that x is positive. (So properly speaking, *Mathematica* should not have found a limit for each of `In[16]` and `In[17]`.)

3. Plot graphs of
$$x \log x \text{ and } x^x$$
on the interval $[0, 2]$ to convince yourself that *Mathematica*'s answers for `In[16]` and `In[17]` are indeed correct for $x \to 0^+$.

4. Consider the function
$$f(x) = x^{1/x}.$$

(a) Plot a graph of f on various intervals of the positive x-axis to see if
$$\lim_{x \to 0^+} f(x) \text{ and } \lim_{x \to \infty} f(x)$$
exist. What does it look like these values should be?

(b) Calculate $f(0.5)$, $f(0.2)$, $f(0.1)$, $f(0.01)$. What does it appear the limit of f as $x \to 0^+$ should be?

(c) Calculate $f(10)$, $f(20)$, $f(50)$, $f(100)$. What does it appear the limit of f as $x \to \infty$ should be?

(d) Use `Limit` to calculate the above two limits? Does *Mathematica* give the right values?

5. Find another example of a limit that exists, but which *Mathematica* cannot evaluate.

2.2 Limits That *Mathematica 1.2* Can't Do

As you should be aware of by now, there are many limits that *Mathematica* does not evaluate correctly with the function `Limit`. This is important to know! In many cases it may well be necessary to cross-check an answer supplied by *Mathematica*.[1] Indeed, in the true scientific spirit, you should be skeptical of *any* result produced by *Mathematica*. However, as you work with it more and more, you will quickly learn to appreciate what it *can* do, and get a feel for the things it *can't* do. This section will show you some kinds of limits that *Mathematica* definitely cannot do. In some of these cases it will inform you that it can't find a limit, but in others it will blithely give you an incorrect answer.

Example 1. Consider the limit
$$\lim_{x \to 0} x \sin 1/x.$$
This is one of those examples in which *Mathematica* informs you that it can't find a limit:

```
In[1]:= f[x_] := x Sin[1/x]
In[2]:= Limit[f[x], x -> 0]
Limit::nlm: Could not find definite limit.
                  1
Out[2]= Limit[x Sin[-], x -> 0]
                  x
```

---
[1] The examples in this edition of the book are all based on *Mathematica 1.2*; things appear better in Version 2.0. Both versions can be setup in parallel on some computers during a changeover.

2.2. LIMITS THAT MATHEMATICA 1.2 CAN'T DO

However, a limit exists — the limit is 0 — and we can prove this simply by using a basic fact about the sine function, namely that

$$-1 \leq \sin\theta \leq 1, \text{ for all } \theta.$$

Thus for all positive x we must have

$$-x \leq x\sin 1/x \leq x,$$

and for all negative x we must have

$$x \leq x\sin 1/x \leq -x.$$

We can plot graphs of $y = \pm x$ and f to see this:

```
In[3]:= Plot[{x, -x, f[x]}, {x, -1, 1}];
```

See Figure 2.2. This shows that $\lim_{x \to 0} f(x) = 0$.

Example 2. *Mathematica* can evaluate limits of a function defined in terms of cases everywhere except at the points dividing one case from the other. Consider the function h defined by

$$h(x) = \begin{cases} x^2 + 1, & \text{if } x \geq 0 \\ x, & \text{otherwise} \end{cases}$$

Let us find limits of h at different values of x:

```
In[4]:= h[x_] := If[x < 0, x, x^2 + 1]
In[5]:= Limit[h[x], x -> 1]
Out[5]= 2
In[6]:= Limit[h[x], x -> -1]
Out[6]= -1
```

These limits are both correct. But the next one is not:

```
In[7]:= Limit[h[x], x -> 0]
Out[7]= 1
```

What *Mathematica* returns is $h(0)$, even though no limit exists. We can plot the graph of h to see what is happening. See Figure 2.2.

```
In[8]:= Plot[h[x], {x, -2, 2}];
```

It is a general feature of *Mathematica*, Version 1.2, that it can not evaluate one-sided limits. No matter how we approach 0, *Mathematica* will always respond with the value of $h(0)$ as a limit:

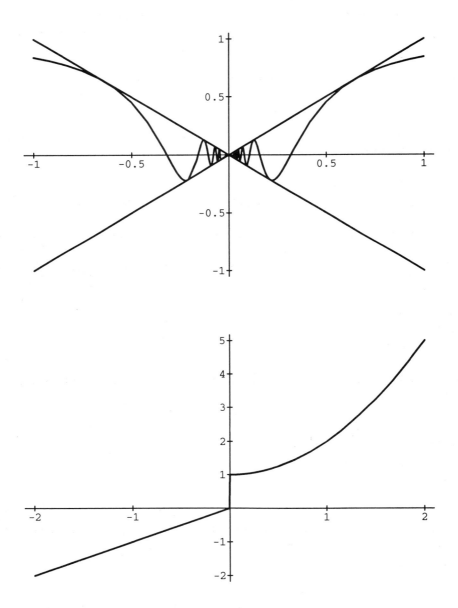

Figure 2.2: Top: $\lim_{x \to 0} x \sin 1/x = 0$. Bottom: *Mathematica* gives a limit for the above function at $x = 0$, even though no limit exists.

2.2. LIMITS THAT MATHEMATICA 1.2 CAN'T DO

```
In[9] := Limit[h[x^2], x -> 0]  (* x^2 always positive;
                                    limit should be 1 *)
Out[9] = 1
In[10] := Limit[h[-x^2], x -> 0]  (* -x^2 always negative;
                                      limit should be 0 *)
Out[10] = 1
```

Example 3. The greatest integer function, $[[x]]$, is defined by

$$[[x]] = n, \text{ if } n \text{ is an integer and } n \leq x < n+1..$$

It gives the greatest integer less than or equal to x; see p 6, Edwards and Penney[3]. In *Mathematica* this function is called `Floor`:

```
In[11] := ?Floor
Floor[x] gives the greatest integer less than or equal to x.
```

Thus we can plot the greatest integer function directly, but note that *Mathematica*, as usual, tries to join separate pieces of this step function with vertical lines. See Figure 2.3. The following input also highlights the points on the graph of $[[x]]$ for which $y = x$.

```
In[11] := Plot[Floor[x], {x, -3, 3}];
In[12] := ListPlot[Table[{i, i}, {i, -3, 3}]];
In[13] := Show[%, %%];
```

As in Example 2, *Mathematica* is able to evaluate the limit as long as x is not approaching an integer:

```
In[14] := Limit[Floor[x], x -> 2.5]
Out[14] = 2
In[15] := Limit[Floor[x], x -> 2]
Out[15] = 2
```

Note that *Mathematica* responds with an answer in `Out[15]`, even though at $x = 2$, and at any other integer, for that matter, no limit exists, since the left–handed and right–handed limits are not the same.

Example 4. Consider the limit

$$\lim_{x \to 0} e^{1/x},$$

which does not exist, since the right–sided limit is obviously infinite. That is, as $x \to 0^+$, $1/x \to \infty$. But the left-sided limit *does* exist. It is equal to 0; you can see this by plotting $e^{1/x}$ on an interval to the left of $x = 0$. See Figure 2.3. The inputs and outputs for this example are as follows:

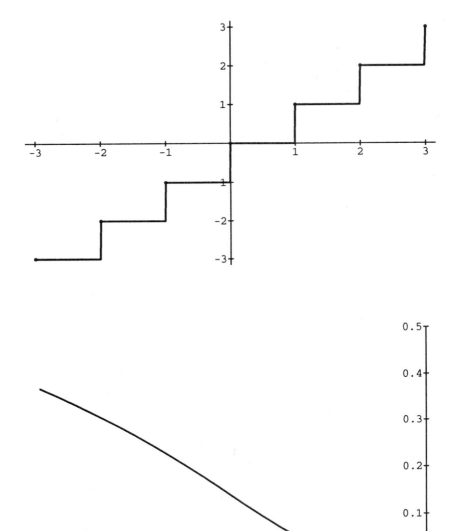

Figure 2.3: Top: graph of $y = [[x]]$; no limit exists at $x = n$, for any integer n. Bottom: $\lim_{x \to 0^-} e^{1/x} = 0$.

2.2. LIMITS THAT MATHEMATICA 1.2 CAN'T DO

```
In[16]:= Limit[Exp[1/x], x -> 0]
Limit::nlm: Could not find definite limit.
                 1/x
Out[16]= Limit[E   , x -> 0]
In[17]:= Plot[Exp[1/x], {x, -1, -0.01}];
In[18]:= Show[%, PlotRange -> {{-1, 0}, {0, 0.5}}];
```

Example 5. Consider the geometric series

$$s(n) = \sum_{i=0}^{n} \frac{1}{2^i}.$$

We can evaluate this sum for any finite value of n directly by using **Sum**. The following table gives the values of $s(n)$ for n between 1 and 10.

```
In[19]:= s[x_] := Sum[1/2^i, {i, 0, x}]
In[20]:= Table[s[n], {n, 1, 10}] // N
Out[20]= {1.5, 1.75, 1.875, 1.9375, 1.96875, 1.98437, 1.99219, 1.99609,

>        1.99805, 1.99902}
```

Note that the values are approaching the limiting value of 2; after all, as you may recall,

$$\lim_{n \to \infty} s(n) = \frac{1}{1 - \frac{1}{2}} = 2.$$

But you cannot obtain this result directly from *Mathematica*:

```
In[21]:= Limit[s[n], n -> Infintiy]
SeriesData::bsd:
   SeriesData encountered arguments
     SeriesData[n, Infintiy, {1. + Infintiy, {0., 0., 0.}}, 0., 3., 1.].
SeriesData::bsd:
   SeriesData encountered arguments
     SeriesData[n, Infintiy, {1. + Infintiy, {0, 0, 0}}, 0, 3., 1.].
              -i                                                      3
Out[21]= Sum[2  , {i + {0, 0, 1} (-Infintiy + n) + O[-Infintiy + n]  ,

                                                             3
>       {0, 0, 1} (-Infintiy + n) + O[-Infintiy + n]  ,

                                                                 3
>       Infintiy + {0, 0, 1} (-Infintiy + n) + O[-Infintiy + n]  }]
```

All we obtain is a list of error messages. However, *Mathematica* does offer another function, **NSum**, specifically for calculating limits of infinite series:

```
In[22]:= NSum[1/2^i, {i, 0, Infinity}]
Out[23]= 2.
```

Here are two more examples of using `NSum` for evaluating infinite series — one in which no limit exists, and one for which there is a limit.

```
In[24]:= NSum[1/i, {i, 1, Infinity}]
NSum::sumdiv: Sum apparently diverges.
                1
Out[24]= NSum[-, {i, 1, Infinity}]
                i
In[25]:= NSum[1/i^2, {i, 1, Infinity}]
Out[25]= 1.64493
```

This last limit is not at all easy to evaluate exactly. *Mathematica* computes a six-digit answer; the precise value is $\pi^2/6$, but this is hardly obvious.

> **MATHEMATICA FUNCTIONS INTRODUCED IN THIS SECTION**
> Floor   NSum

Exercises:

1. Repeat the analysis of Example 1 for the following two limits. Does *Mathematica* give the correct result in either case?

   (a) $\lim_{x \to 0} x^2 \sin 1/x$

   (b) $\lim_{x \to 0} \sin 1/x$

2. Consider the function f such that
$$f(x) = \frac{|x|}{x}.$$

   (a) Is there a limit as $x \to 0$?

   (b) What about left– and right–sided limits at $x = 0$?

   (c) What does *Mathematica* return as $\lim_{x \to 0} f(x)$? Do you have any idea what is behind its answer?

3. Consider the function $h(x) = x[[x]]$.

   (a) Plot the graph of h on the interval $[-2, 2]$, both with *Mathematica*, and *correctly* with paper and pencil.

(b) What is $\lim_{x\to 0} h(x)$?

(c) Does *Mathematica* give the correct answer for the limit of part b)?

(d) Are there any other integers n among $\pm 1, \pm 2$ for which the limit of h as $x \to n$ exists?

(e) What does *Mathematica* return for the limits in part d)?

4. Define a function f such that

$$f(x) = \frac{3^x - 3^{-x}}{3^x + 3^{-x}}.$$

(a) What is $\lim_{x\to\infty} f(x)$?

(b) What is $\lim_{x\to-\infty} f(x)$?

(c) Does *Mathematica* give the correct answers for the limits in parts a) and b)?

5. Define a function g such that

$$g(x) = \frac{1 - 2^{1/x}}{1 + 2^{1/x}}.$$

(a) What is $\lim_{x\to\infty} g(x)$?

(b) What is $\lim_{x\to-\infty} g(x)$?

(c) Does *Mathematica* give the correct answers for the limits in parts a) and b)?

(d) What does *Mathematica* find for $\lim_{x\to 0} g(x)$?

(e) Find both one-sided limits of g at $x = 0$.

6. Define a function k such that

$$k(x) = \frac{1}{3^{1/x} + 1}.$$

(a) Find both one-sided limits of k at $x = 0$.

(b) Find both one-sided limits of $xk(x)$ at $x = 0$.

(c) What does *Mathematica* return as

$$\lim_{x\to 0} xk(x)?$$

7. NSum can be used for finite sums as well. Calculate the sum of the first 10 terms of the following series using NSum:

(a) $1 + 2 + 3 + \cdots + n + \cdots$

(b) $1^2 + 2^2 + 3^2 + \cdots + n^2 + \cdots$

(c) $\sin \pi/4 + \sin 2\pi/4 + \sin 3\pi/4 + \cdots + \sin n\pi/4 + \cdots$

8. Use `NSum` to see if *Mathematica* can find the sum of the following infinite series:

(a) $\dfrac{1}{1^4} + \dfrac{1}{2^4} + \dfrac{1}{3^4} + \dfrac{1}{4^4} + \cdots = \dfrac{\pi^4}{90}$

(b) $\dfrac{1}{1^6} + \dfrac{1}{2^6} + \dfrac{1}{3^6} + \dfrac{1}{4^6} + \cdots = \dfrac{\pi^6}{945}$

(c) $\dfrac{1}{1^2} - \dfrac{1}{2^2} + \dfrac{1}{3^2} - \dfrac{1}{4^2} + \cdots = \dfrac{\pi^2}{12}$

(d) $1 - \dfrac{1}{3} + \dfrac{1}{5} - \dfrac{1}{7} + \dfrac{1}{9} + \cdots = \dfrac{\pi}{4}$

(e) $\dfrac{1}{1 \cdot 3} + \dfrac{1}{3 \cdot 5} + \dfrac{1}{5 \cdot 7} + \dfrac{1}{7 \cdot 9} + \cdots = \dfrac{1}{2}$

2.3 Continuity

A function f is said to be *continuous at* $x = a$ if

- $x = a$ is in the domain of f,
- $\lim_{x \to a} f(x)$ exists, and
- $\lim_{x \to a} f(x) = f(a)$.

A function f is said to be *continuous on an interval* $[a, b]$ if it is continuous at every point in the interval.[2] If a function is *not* continuous at $x = a$, then it is said to be *discontinuous* at $x = a$, and $x = a$ is called a *discontinuity* of f. It is at discontinuities that *Mathematica* may have difficulty evaluating certain limits, as the examples from Sections 2.1 and 2.2 show. In fact, on p 436, Wolfram[8], it says:

> If you give `Limit` a function it does not know, it assumes that the function is continuous at the limit point.

This explains *Mathematica*'s response in Examples 2 and 3 of Section 2.3, but it is not the full story. Sometimes, *even if* the function is continuous, *Mathematica* will still not find the limit. See Exercises 1 and 2 below.

---

[2] At the left–end point, $x = a$, we only require that $\lim_{x \to a^+} f(x) = f(a)$; at the right–end point, $x = b$, we only require that $\lim_{x \to b^-} f(x) = f(b)$.

2.3. CONTINUITY

Example 1. It is also at discontinuities that *Mathematica* has difficulty plotting a function. Consider the function k defined by

$$k(x) = \frac{|x-3|}{x-3}.$$

Suppose we ask *Mathematica* to plot the graph of k on the interval $[0,4]$, which includes the only discontinuity of k, namely $x = 3$.

```
In[1]:= k[x_] := Abs[x - 3]/(x - 3)
In[2]:= Plot[k[x], {x, 0, 4}];
```

Everything is correct about the graph of k as *Mathematica* plots it *except* for the vertical line segment at $x = 3$. What we see here is *Mathematica's* attempt to plot k as if it were continuous on $[0,4]$. Continuous functions on a closed interval have the Intermediate Value Property, see p 62, Edwards and Penney[3], and it is precisely this property that *Mathematica's* graph of k has, even though it is not a continuous function. To obtain the correct graph of a discontinuous function using *Mathematica* it is usually necessary to plot the function on separate intervals which do not include the dicontinuities, except possibly as end points, and then to combine the separate pieces of the graph with `Show`.

```
In[3]:= Plot[k[x], {x, 0, 3}];
                              1
Power::infy: Infinite expression -- encountered.
                              0.
Infinity::indt: Indeterminate expression 0. ComplexInfinity encountered.
Plot::notnum: k[x] does not evaluate to a real number at x=3..
In[4]:= Plot[k[x], {x, 3, 4}];
                              1
Power::infy: Infinite expression -- encountered.
                              0.
Infinity::indt: Indeterminate expression 0. ComplexInfinity encountered.
Plot::notnum: k[x] does not evaluate to a real number at x=3..
In[5]:= Show[%%, %];
```

This is how the graph of k was obtained for the top of Figure 2.4.

Example 2. When infinite discontinuities (ie. vertical asymptotes) are involved, *Mathematica* may or may not include extraneous vertical lines in its plots. For instance, consider the function f defined by

$$f(x) = \frac{x^2 + 1}{1 - x^2}$$

which has infinite discontinuities at $x = \pm 1$. On one plot range an extraneous vertical line shows up; on the other, no such lines show up:

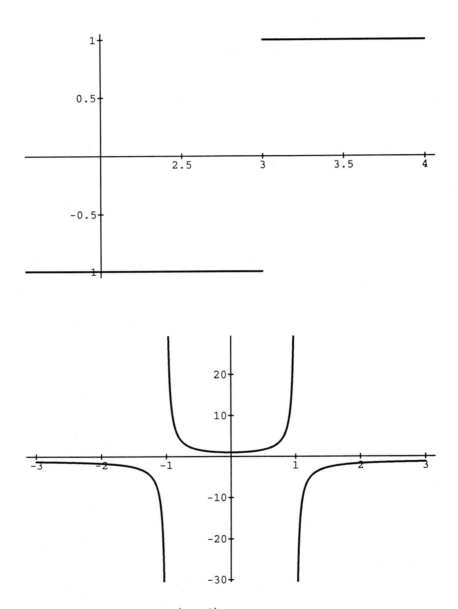

Figure 2.4: Top: graph of $y = \dfrac{|x-3|}{x-3}$ has a jump discontinuity at $x = 3$. Bottom: The graph of $y = \dfrac{x^2+1}{1-x^2}$ has infinite discontinuities at $x = \pm 1$.

2.3. CONTINUITY

```
In[6] := f[x_] := (x^2 + 1)/(1 - x^2)
In[7] := Plot[f[x], {x, -2, 2}];
```
$$1$$
Power::infy: Infinite expression -- encountered.
$$0.$$
Plot::notnum: f[x] does not evaluate to a real number at x=1..
```
In[8] := Plot[f[x], {x, -3, 3}];
```
$$1$$
Power::infy: Infinite expression -- encountered.
$$0.$$
Plot::notnum: f[x] does not evaluate to a real number at x=-1..
$$1$$
Power::infy: Infinite expression -- encountered.
$$0.$$
Plot::notnum: f[x] does not evaluate to a real number at x=1..

On the interval $[-3, 3]$, *Mathematica* produces a completely correct graph. Your most recent output should look like the bottom of Figure 2.4.

Example 3. If a function is continuous, then the plots produced by *Mathematica* should be correct and useful — as long as you plot the function on a suitable domain, as pointed out in Section 1.7. One type of problem for which *Mathematica's* plotting is very useful is the isolation of roots, cf. Example 7, p 63, Edwards and Penney[3]. Here we shall do a similar analysis for the equation

$$x^3 + 4x^2 - 6 = 0.$$

Let $g(x) = x^3 + 4x^2 - 6$. Since g is everywhere continuous, we could use the Intermediate Value Property to find consecutive integers between which each real root of the equation $g(x) = 0$ must lie. But it is much easier simply to plot g and look at its graph to see between which integers each root lies. See Figure 2.5.

```
In[9] := g[x_] := x^3 + 4x^2 - 6
In[10] := Plot[g[x], {x, -5, 2}];
```

From the graph, it is clear that the equation $x^3 + 4x^2 - 6 = 0$, has three real roots, one in the interval $[-4, -3]$, a second one in the interval $[-2, -1]$, and a third root in the interval $[1, 2]$. Compare this graphical approach with the direct computational approach of showing that g has different signs at each end of the above three intervals; graphing is just more convenient.

```
In[11] := {g[-4], g[-3]}
Out[11]= {-6, 3}
In[12] := {g[-2], g[-1]}
```

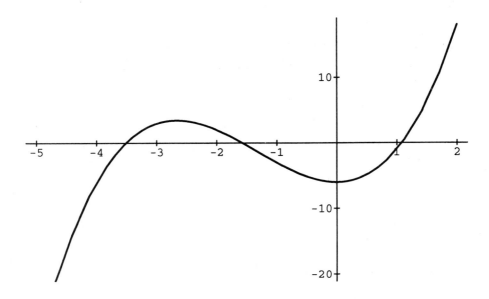

Figure 2.5: Plotting the graph of $y = x^3 + 4x^2 - 6$ to see how many real roots there are to the equation $x^3 + 4x^2 - 6 = 0$..

```
Out[12]= {2, -3}
In[13]:= {g[1], g[2]}
Out[13]= {-1, 18}
```

Exercises:

1. Define a function f such that
$$f(x) = \begin{cases} 0, & \text{if } x = 0 \\ x \sin 1/x, & \text{otherwise} \end{cases}$$

   (a) Is f continuous at $x = 0$?

   (b) Use `Limit` to evaluate $\lim_{x \to 0} f(x)$. Does *Mathematica* give the correct answer?

2. Define a function g such that
$$g(x) = \frac{x-1}{1-|x|}.$$

   (a) Find all the discontinuities of g.

2.4. OPTIONAL: THE DEFINITION OF A LIMIT

(b) Plot the graph of g using *Mathematica*.

(c) Use `Limit` to find the limits of g at its discontinuities.

(d) Does either limit from part c) actually exist?

(e) At which of the two discontinuities of g is it possible to redefine g so that it becomes continuous?

(f) Redefine g in *Mathematica* using `If` so that it becomes continuous at $x = 1$. Does *Mathematica* now give the correct limit for g as $x \to 1$?

3. For each of the following functions f determine all discontinuities, and then, by constructing its graph piece by piece if necessary, use *Mathematica* to plot a correct graph of f.

(a) $f(x) = \dfrac{x}{x^2 - 1}$

(b) $f(x) = \dfrac{x - 2}{x^2 - 4}$

(c) $f(x) = \dfrac{x + 1}{x^2 - x - 6}$

(d) $f(x) = x - [[x]]$ on $[-3, 3]$

(e) $f(x) = \begin{cases} -x, & \text{if } x \leq 0 \\ x^2, & \text{if } x > 0 \end{cases}$

(f) $f(x) = \begin{cases} x + 1, & \text{if } x < 0 \\ 3 - x, & \text{if } x \geq 0 \end{cases}$

(g) $f(x) = [[x^2/3]]$ on $[-3, 3]$

4. For the following equations, find consecutive integers between which each root lies:

(a) $x^3 + x + 1 = 0$

(b) $x^3 - 3x^2 + 1 = 0$

(c) $x^4 + 2x - 4 = 0$

(d) $x^5 - 5x^3 + 3 = 0$

(e) $\sin x + x - 2 = 0$

(f) $2^{x+1} - 2x^2 - 1 = 0$

2.4 Optional: The Definition of a Limit

The formal definition of
$$\lim_{x \to a} f(x) = L$$
is the following:

for every $\epsilon > 0$ there is a $\delta > 0$ such that

$$(2.1) \qquad 0 < |x - a| < \delta \Rightarrow |f(x) - L| < \epsilon.$$

This somewhat intimidating definition can easily be rendered into *Mathematica* as follows:

for every $\epsilon > 0$ there is a $\delta > 0$ such that if f is plotted on the interval $[a - \delta, a + \delta]$, then the range of f should be in the interval $[L - \epsilon, L + \epsilon]$.

A function for carrying out such a graphical test of the limit definition is defined in the pacakage `Chap2.m`; it is called `epsilondeltatest`:

```
In[1]:= << Chap2.m    (* Out[1] will be a usage statement *)
In[2]:= ?epsilondeltatest
epsilondeltatest[f,{a,d},{l,e}] plots a graph of f on the interval
{a -d,a+d} to see if |f[x]-l|<e for all x such that |x-a|<d.
```

Note that in this function the traditional Greek letters ϵ and δ have been replaced by e and d. In this section we will use the defined function `epsilondeltatest` to investigate the existence or non–existence of limits in terms of the formal definition. These examples are not meant to constitute *proof* of a limit — although we can certainly *disprove* a limit with `epsilondeltatest` — but are meant simply to familiarize the student with the ideas involved in the formal definition.

Example 1. Let $f(x) = x^2$; we shall consider the limit of f as $x \to 1$. We can calculate the limit with `Limit` to obtain the (obvious) result:

```
In[2]:= f[x_] := x^2
In[3]:= Limit[f[x], x -> 1]
Out[3]= 1
```

What is involved in trying to prove this limit formally? Take a concrete example: let $\epsilon = 0.1$. We must find a δ such that condition (2.1) is met. The following inputs test $\delta = 0.1, 0.05, 0.04$ and 0.03 in turn. See Figure 2.6

```
In[4]:= epsilondeltatest[f, {1, 0.1}, {1, 0.1}];
In[5]:= epsilondeltatest[f, {1, 0.05}, {1, 0.1}];
In[6]:= epsilondeltatest[f, {1, 0.04}, {1, 0.1}];
In[7]:= epsilondeltatest[f, {1, 0.03}, {1, 0.1}];
```

As you can see from the above graphs, once you have found one δ that meets the requirements of condition (2.1), any smaller choice will do.

Now suppose we decrease ϵ to 0.01. As you can check below, $\delta = 0.007$ is too large, but $\delta = 0.004$ (and anything smaller) is OK. See Figure 2.7

2.4. OPTIONAL: THE DEFINITION OF A LIMIT

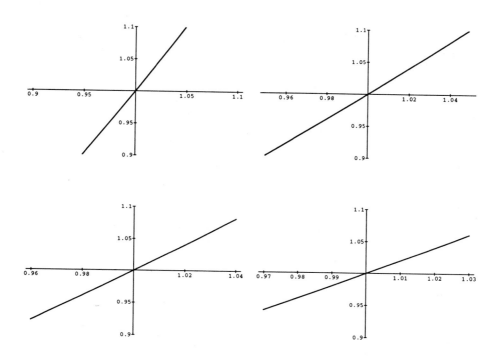

Figure 2.6: $\lim_{x \to 1} x^2 = 1$. For $\epsilon = 0.1$ four choices of δ are illustrated: $\delta = 0.1$ is too large (top left); $\delta = 0.05$ is slightly too large (top right); $\delta = 0.4$ (bottom left) or 0.3 (bottom right) are small enough.

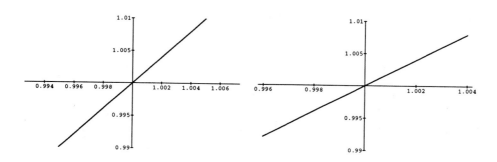

Figure 2.7: $\lim_{x \to 1} x^2 = 1$. For $\epsilon = 0.01$, $\delta = 0.007$ is too large (left) but $\delta = 0.004$ is small enough.

```
In[8]:= epsilondeltatest[f, {1, 0.007}, {1, 0.01}];
In[9]:= epsilondeltatest[f, {1, 0.004}, {1, 0.01}];
```

As you can see from Figures 2.6 and 2.7 decreasing ϵ is a way of "zooming in" on the function near the point $x = 1$.

Example 2. Let us now consider a more complicated case:

$$\lim_{x \to 0} x \sin 1/x.$$

Mathematica cannot evaluate this limit; but we know the limit is 0, as we determined in Example 1 of Section 2.2. Suppose however, this limit is unknown to you, and for some reason you suspect that

$$\lim_{x \to 0} x \sin 1/x = 1.$$

We can show this is impossible by using `epsilondeltatest` with $\epsilon = 0.1$ and taking $\delta = 0.1$. See Figure 2.8.

```
In[10]:= f[x_] := x Sin[1/x]
In[11]:= epsilondeltatest[f, {0, 0.1}, {1, 0.1}];
```

What we see is that none of the graph of f is in the picture of `Out[11]`. This would also occur for any choice of δ smaller than $\delta = 0.1$; thus condition (2.1) cannot be satisfied for any δ if $\epsilon = 0.1$, so the limit cannot be 1. Of course things work much better if we try to show the limit is 0. In fact, as the following inputs show, we can take $\delta = \epsilon$. (See Figure 2.8.) Can you prove this in general?

```
In[12]:= epsilondeltatest[f, {0, 0.1}, {0, 0.1}];
In[13]:= epsilondeltatest[f, {0, 0.01}, {0, 0.01}];
In[14]:= epsilondeltatest[f, {0, 0.1}, {0, 0.01}];
```

Example 3. As a last example, consider the function h of Example 2, Section 2.2 defined by

$$h(x) = \begin{cases} x^2 + 1, & \text{if } x \geq 0 \\ x, & \text{otherwise} \end{cases}$$

This function is not continuous at $x = 0$, because $\lim_{x \to 0} h(x)$ does not exist. We can show this with `epsilondeltatest` as follows. Suppose you think the limit as $x \to 0$ is 1, which is *Mathematica's* answer. You can show this is not true by taking $\epsilon = 0.1$ and $\delta = 0.1$. What you see, Figure 2.9, is that the left half of the graph of h is outside the required plot range. If you use the same choices of ϵ and δ, but with $L = 0$, you find the right half of the graph is outside the required plot range. See Figure 2.9.

2.4. OPTIONAL: THE DEFINITION OF A LIMIT

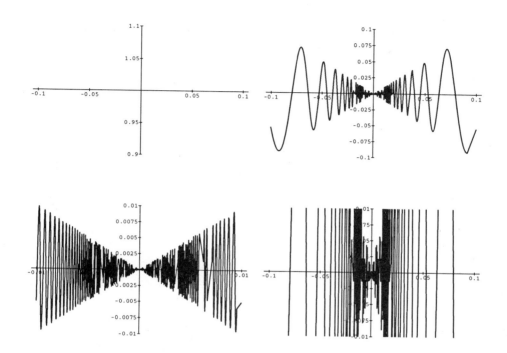

Figure 2.8: Top left: $\lim_{x \to 0} x \sin 1/x \neq 1$. The limit is 0: if $\epsilon = 0.1$ take $\delta = 0.1$ (top right); if $\epsilon = 0.01$, $\delta = 0.1$ is too large (bottom right) but $\delta = 0.01$ is small enough (bottom left).

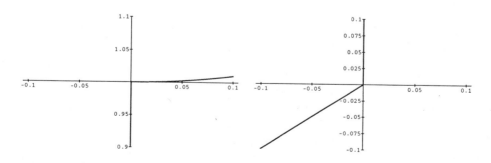

Figure 2.9: $h(x) = x$, if $x < 0$; $h(x) = x^2 + 1$, otherwise. Left: $\lim_{x \to 0} h(x) \neq 1$; right: nor is it 0.

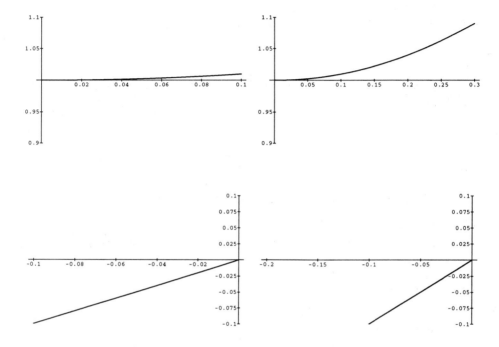

Figure 2.10: Top: $\lim_{x \to 0^+} h(x) = 1$. Bottom: $\lim_{x \to 0^-} h(x) = 0$.

```
In[15]:= h[x_] := If[x < 0, x, x^2 + 1]
In[16]:= epsilondeltatest[h, {0, 0.1}, {1, 0.1}];
In[17]:= epsilondeltatest[h, {0, 0.1}, {0, 0.1}];
```

Thus the limit of h as $x \to 0$ can be neither 1 or 0. But the left– and right–sided limits do exist for h at $x = 0$. To formally analyse one–sided limits there are two variations of epsilondeltatest defined in the package Chap2.m: repsilondeltatest is for right–sided limits, lepsilondeltatest is for left–sided limits. These functions restrict the plotting of f to one side of the point $x = a$. Of the four following inputs, the first two show that for

$$\lim_{x \to 0^+} h(x) = 1$$

if $\epsilon = 0.1$, $\delta = 0.1$ or even 0.3 are sufficiently small. The last two show that for

$$\lim_{x \to 0^-} h(x) = 0$$

if $\epsilon = 0.1$, $\delta = 0.2$ is too large, but $\delta = 0.1$ is sufficiently small. See Figure 2.10.

2.4. OPTIONAL: THE DEFINITION OF A LIMIT

```
In[18] := repsilondeltatest[h, {0, 0.1}, {1, 0.1}];
In[19] := repsilondeltatest[h, {0, 0.3}, {1, 0.1}];
In[20] := lepsilondeltatest[h, {0, 0.1}, {0, 0.1}];
In[21] := lepsilondeltatest[h, {0, 0.2}, {0, 0.1}];
```

FUNCTIONS CONTAINED IN THE PACKAGE Chap2.m		
epsilondeltatest	repsilondeltatest	lepsilondeltatest

Exercises:

1. Consider $\lim_{x \to 0} |x| = 0$.

    (a) Find the largest δ which satisfies condition (2.1) if $\epsilon = 0.1$.

    (b) Repeat part a) for $\epsilon = 0.01$.

    (c) Make a conjecture as to how large δ can be in terms of ϵ. Can you prove your conjecture?

2. Repeat Exercise 1 for $\lim_{x \to 1} x^3 = 1$.

3. Repeat Exercise 1 for $\lim_{x \to 2} \frac{x^2 - 4}{x^2 + x - 6} = 0.8$.

4. Let $f(x) = 1/x$. Show that $\lim_{x \to 0} f(x) \neq 1$.

5. Let $f(x) = |x|/x$. Show that $\lim_{x \to 0} f(x) \neq -1$.

6. Let $f(x) = \sin 1/x$. Show that $\lim_{x \to 0} f(x) \neq 0.5$. Can you generalize your argument to show that f has no limit as $x \to 0$?

7. Let $f(x) = \dfrac{3}{1 + 4^{1/x}}$.

    (a) What are the one-sided limits of f at $x = 0$?

    (b) Use `repsilondeltatest` and `lepsilondeltatest` to find the largest δ for each the left and right sided limits of f at $x = 0$ if $\epsilon = 0.05$.

8. The package Chap2.m does not include any functions for dealing formally with limits as $x \to \pm\infty$. This is not necessary in the sense that such limits can be calculated in terms of one-sided limits as follows:

    (a) Show that
    $$\lim_{x \to \infty} f(x) = L \Leftrightarrow \lim_{x \to 0^+} f(1/x) = L.$$

(b) What is the corresponding result for limits as $x \to -\infty$?

9. For each of the following, find the indicated limit, L, and find N such that
$$x > N \Rightarrow |f(x) - L| < \epsilon,$$
for the indicated value of ϵ.

(a) $L = \lim\limits_{x \to \infty} \dfrac{x^2 + 1}{1 - x^2}$; $\epsilon = 0.1$

(b) $L = \lim\limits_{x \to \infty} \dfrac{\cos x}{x}$; $\epsilon = 0.1$

(c) $L = \lim\limits_{x \to \infty} x e^{-x}$; $\epsilon = 0.1$

(d) $L = \lim\limits_{x \to \infty} x - \sqrt{x^2 + 3x}$; $\epsilon = 0.01$

(e) $L = \lim\limits_{x \to \infty} \dfrac{x}{\sqrt{x^2 + 1}}$; $\epsilon = 0.01$

Chapter 3

The Derivative

> *I sought a Method of determining Quantities from the Velocities of the Motions or Increments, with which they are generated; and calling these Velocities of the Motions or Increments* Fluxions, *and the generated Quantities* Fluents, *I fell by degrees upon the Method of Fluxions, which I have made use of here in the Quadrature of Curves, in the Years 1665 and 1666.*—I.Newton, Tractatus de Quadratura Curvarum 1704

3.1 The Derivative as Slope of the Tangent

The derivative of a function is so named because it is *derived* from the definition of a given function and nothing else. What it does is most clearly understood as a geometric process: it provides at each point the *best linear approximation* to the function. This concept was out of reach of the superb geometers of ancient Greece because they did not have the concept of a limiting value at a point, and that is essential for the approximation to be shown to be *best*.

The tangent to f at a point $x = a$ is defined as the limiting position of secants joining the points $(a, f(a))$ and $(a+h, f(a+h))$, as $h \to 0$. To illustrate this concept we will take $f(x) = \sin x$ and find its tangent at $x = 1$. There is a defined function, secant, in the package Chap3.m which will plot the secant lines for us:

```
In[1]:=<< Chap3.m
In[2]:= ?secant
secant[f, a][p] plots f and the secant joining (a, f[a]) and (p, f[p]).
```

Example 1. We can now proceed to plot many secants to f at $x = 1$. If you combine some of the following graphics outputs (with Show), you should produce a picture something like the top part of Figure 3.1.

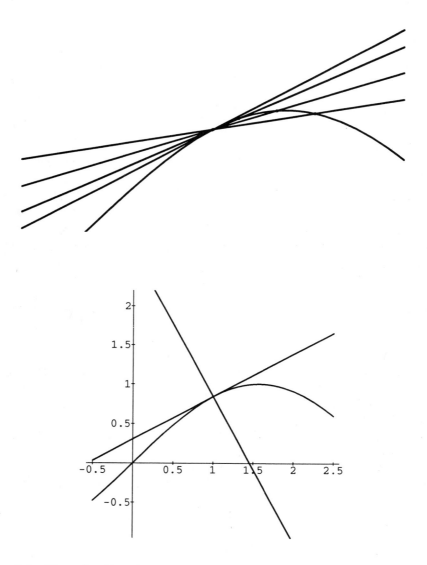

Figure 3.1: Top: family of secants to a curve passing through a common point. Bottom: normal and tangent to $y = \sin x$ at $x = 1$.

3.1. THE DERIVATIVE AS SLOPE OF THE TANGENT

```
In[2]:= secant[Sin, 1][0.3];
In[3]:= secant[Sin, 1][0.5];
In[4]:= secant[Sin, 1][0.7];
In[5]:= secant[Sin, 1][0.9];
In[6]:= secant[Sin, 1][0.95];
In[7]:= secant[Sin, 1][0.99];
In[8]:= secant[Sin, 1][0.999];
```

The tangent to f is the limiting position of these secant lines; in the Appendix there is an animation file, *secant.ani*, which you can view on DOS or NeXT platforms to see the secants in motion as the one point on the curve approaches the other one. The slope of the tangent is defined as the limit of the slopes of the secants as $h \to 0$. This means evaluating

$$\lim_{h \to 0} \frac{f(a+h) - f(a)}{h}.$$

If this limit exists it is called the *derivative of f at $x = a$*, and we denote it by $f'(a)$.

Example 2. For our example of $f(x) = \sin x$, we can calculate the slopes of some of the secants plotted above by using `Table` to generate a list of data:

```
In[9]:= Table[(Sin[1.0 + h] - Sin[1.0])/h, {h, -0.75, -0.05, 0.05}]
Out[9]= {0.792089, 0.77993, 0.767036, 0.753421, 0.739101, 0.724091,
0.708408, 0.692071, 0.675099, 0.657511, 0.639329, 0.620574, 0.601271,
0.581441, 0.56111}
```

Of course, all this data does not prove anything about the actual value of

$$\lim_{h \to 0} \frac{\sin(1+h) - \sin 1}{h}$$

but we can use *Mathematica* to calculate it, both as a limit and as a derivative:

```
In[10]:= Limit[(Sin[1.0 + h] - Sin[1.0])/h, h -> 0]
Out[10]= 0.540302     (* an approximation to Cos[1] = Sin'[1] *)
In[11]:= Sin'[1.0]    (* output will be a numerical approximation *)
Out[11]= 0.540302
```

Once the slope of the tangent to f at $x = a$ is known, then it is a simple matter of using the slope–point form of a line to find its equation,

$$y = f'(a)(x - a) + f(a).$$

This formula is defined in `Chap3.m` as `tangent`:

```
In[12]:= ?tangent
tangent
tangent/: tangent[f_, a_][x_] := f'[a] (x - a) + f[a]
```

Example 3. Thus the tangent to sine at $x = 1$ is given by:

```
In[12]:= tangent[Sin, 1.0][x] // Simplify
Out[12]= 0.301169 + 0.540302 x
```

The line perpendiular to the tangent at $x = a$ is called the *normal* line to f at $x = a$. Since its slope must be the negative reciprocal of the slope of the tangent, it is straightforward to find its equation, too. Thus a formula for the normal to f at $x = a$ is also defined in the package Chap3.m:

```
In[13]:= ?normal
normal
                                  a - x
normal/:  normal[f_, a_][x_]  :=  ----- + f[a]
                                  f'[a]
```

Note that this formula is not well defined if $f'(a) = 0$, in which case the normal is actually a vertical line.

Example 4. Thus the normal to sine at $x = 1$ is given by:

```
In[13]:= normal[Sin, 1.0][x] // Simplify
Out[13]= 2.69229 - 1.85082 x
```

Two other useful functions defined in the package Chap3.m are tangentplot and normalplot:

```
In[14]:= ?tangentplot
tangentplot
tangentplot/: tangentplot[f_, a_, {b_, c_}] :=

>     Plot[{f[x], tangent[f, a][x]}, {x, b, c}]

In[14]:= ?normalplot
normalplot
normalplot/: normalplot[f_, a_, {b_, c_}] :=

>     Plot[{f[x], normal[f, a][x]}, {x, b, c}, AspectRatio -> Automatic]
```

Thus we can use these functions to plot f together with its tangent or normal at $x = a$. Note that normalplot sets the aspect ratio at automatic; this makes the normal *look* perpendicular to the curve.

Example 5. As an example, we can plot the normal and tangent to sine at $x = 1$:

3.1. THE DERIVATIVE AS SLOPE OF THE TANGENT

```
In[14]:= tangentplot[Sin, 1, {-0.5, 2.5}];
In[15]:= normalplot[Sin, 1, {-0.5, 2.5}];
In[16]:= Show[%, %%];
```

This last plot combines the previous two into one, showing both the normal and the tangent; your plot should look something like the picture on the bottom of Figure 3.1.

Finally, if the slope of the tangent to f at $x = a$ is $f'(a)$, then the angle the tangent makes to the x-axis is given by

$$\arctan f'(a)$$

which *Mathematica* will evaluate in radians.

Example 6. Thus we can calculate the angle that the tangent to sine at $x = 1$ makes with the x-axis as follows:

```
In[17]:= ArcTan[Sin'[1]] // N
Out[17]= 0.495367
```

Exercises:

1. As in Example 1, plot a few secants to the following functions, f, to include the indicated point, $x = a$. Pick at least 5 on each side of the indicated point. Then combine all your graphs into one with Show. Is the function differentiable at the indicated point?

    (a) $f(x) = x^2$; at $a = 1$
    (b) $f(x) = \log x$; at $a = 2$
    (c) $f(x) = \sqrt{9 + x^2}$; at $a = 4$
    (d) $f(x) = |x|$; at $a = 0$
    (e) $f(x) = x^{2/3}$; at $a = 0$
    (f) $f(x) = \begin{cases} x^2 \sin 1/x, & \text{if } x \neq 0 \\ 0, & \text{otherwise} \end{cases}$; at $a = 0$
    (g) $f(x) = x^{1/3}$; at $a = 0$

2. For those functions, f, of Exercise 1 which are differentiable at the indicated point, $x = a$, plot both the tangents and the normals to f at $x = a$.

3. Consider the function f defined by $f(x) = |x^2 - 1|$.

    (a) Plot f on the interval $[-2, 2]$.
    (b) At which points does it appear as if f is not differentiable?

(c) Plot secants to f around each of the three points $x = -1$, $x = 0$, and $x = 1$.

(d) Is f differentiable at $x = \pm 1$?

3.2 Differentiation with *Mathematica*

The definition of the derivative and the basic rules of differentiation are built in to *Mathmatica*. Thus:

```
In[1]:= Limit[(f[a + h] - f[a])/h, h -> 0]   (* definition *)
Out[1]= f'[a]
In[2]:= D[f[x] + g[x], x]   (* derivative of a sum *)
Out[2]= f'[x] + g'[x]
In[3]:= D[f[x] g[x], x]   (* product rule *)
Out[3]= g[x] f'[x] + f[x] g'[x]
In[4]:= D[f[g[x]], x]   (* chain rule *)
Out[4]= f'[g[x]] g'[x]
In[5]:= D[f[x]/g[x], x]   (* quotient rule *)
         f'[x]      f[x] g'[x]
Out[5]=  -----   -  ----------
         g[x]           2
                      g[x]
```

This last formula may not look like the quotient rule as you remember it; but if you get a common denominator for Out[5] you should see the familiar formula:

```
In[6]:= Together[%]
        g[x] f'[x] - f[x] g'[x]
Out[6]= -----------------------
                   2
                 g[x]
```

Note that *Mathematica* gives its answers in terms of functional notation, as opposed to Leibnizian notation, or differential notation, etc. You should become thoroughly familiar with functional notation — especially in terms of the chain rule. Consider the following example:

Example 1. There is an important difference between

$$\frac{df(x^2 + 3\sin x)}{dx} \text{ and } f'(x^2 + 3\sin x).$$

Mathematica knows it, and you should too.

3.2. DIFFERENTIATION WITH MATHEMATICA

```
In[7]:= D[f[x^2 + 3 Sin[x]], x]
Out[7]= (2 x + 3 Cos[x]) f'[x  + 3 Sin[x]]
                                 2
In[8]:= f'[x^2 + 3 Sin[x]]
Out[8]= f'[x  + 3 Sin[x]]
            2
```

If you do not understand the difference between Out[7] and Out[8] you had better review functional notation and the chain rule. Here is a more concrete example which may help you. Let $f(x) = \sqrt{x}$. What is

$$f'(x^2 + 3\sin x)?$$

Write down your answer, and then check it with *Mathematica*:

```
In[9]:= Sqrt'[x^2 + 3 Sin[x]]
                   1
Out[9]= ---------------------
              2
         2 Sqrt[x  + 3 Sin[x]]
```

Although *Mathematica* can differentiate most functions that you will encounter, there are certain types of functions that it can not differentiate. Not surprisingly, these include many of the functions for which *Mathematica* has difficulty evaluating limits. Here are four examples. (See Figure 3.2.)

Example 2. Consider the absolute value function, $|x|$, which is differentiable everywhere except at $x = 0$. However, *Mathematica* cannot differentiate it anywhere:

```
In[10]:= Abs'[x]
Out[10]= Abs'[x]
In[11]:= Plot[Abs[x], {x, -3, 3}];
In[12]:= Abs'[3]
Out[12]= Abs'[3]
```

As you can see from the graph of $y = |x|$, Figure 3.2, the derivative at $x = 3$ should be 1.

Example 3. *Mathematica* cannot differentiate any function defined in terms of If. Consider h defined by

$$h(x) = \begin{cases} x^2 + 1, & \text{if } x \geq 0 \\ x, & \text{otherwise} \end{cases}$$

The derivative of h is

$$h'(x) = \begin{cases} 2x, & \text{if } x > 0 \\ 1, & \text{if } x < 0 \end{cases}$$

See Figure 3.2. Defining h in *Mathematica*, and then differentiating we obtain:

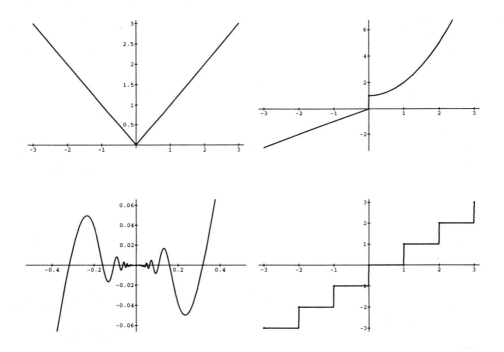

Figure 3.2: Functions which *Mathematica* cannot differentiate: it cannot differentiate $|x|$ (top left), $h(x)$ (top right), or $[[x]]$ (bottom right) anywhere. It can differentiate $x^2 \sin 1/x$ (bottom left) everywhere, except at $x = 0$.

3.2. DIFFERENTIATION WITH MATHEMATICA

```
In[13]:= h[x_] := If[x < 0, x, x^2 + 1]
In[14]:= Plot[h[x], {x, -3, 3}];
In[15]:= D[h[x], x]
              (0,0,1)                    2         (0,1,0)                 2
Out[15]= 2 x If       [x < 0, x, 1 + x ] + If       [x < 0, x, 1 + x ]

           (1,0)             (1,0,0)              2
    +  Less    [x, 0] If           [x < 0, x, 1 + x ]
```

This is a mysterious output — *Mathematica* is considering If as a function of *three* variables, and is trying to calculate its total derivative in terms of partial derivatives. None of which makes any sense to you, probably. Let us just say that *Mathematica* cannot differentiate h. You might think it is possible to circumvent this problem by redefining your functions, or performing the algebra in a different way. However, none of the obvious alternatives work if we insist on dealing with the whole domain:

```
In[16]:= If[x < 0, D[x, x], D[x^2 + 1, x]]
                                    2
Out[16]= If[x < 0, D[x, x], D[x  + 1, x]]
In[17]:= Out[15] /. {x -> -2, x -> 1}
General::bvar: -2 is a number which cannot be used as a variable.
Out[17]= D[-2, -2]
In[18]:= hprime[x_] := If[x < 0, D[x, x], D[x^2 + 1, x]]
In[19]:= hprime[-3]
General::bvar: -3 is a number which cannot be used as a variable.
Out[19]= D[-3, -3]
```

Of course we can deal with the different parts of the function separately and then fit their derivatives together afterwards, being careful to check their existence and definition at the endpoints of the subdomains.

Example 4. Consider the function $f(x) = x^2 \sin 1/x$. *Mathematica* can differentiate f everywhere except at $x = 0$:

```
In[20]:= f[x_] := x^2 Sin[1/x]
In[21]:= f'[x]
              1           1
Out[21]= -Cos[-] + 2 x Sin[-]
              x           x

In[22]:= f'[0]
                            1
Power::infy: Infinite expression - encountered.
                            0
```

```
                                  1
Power::infy: Infinite expression - encountered.
                                  0
Out[22]= -Cos[ComplexInfinity]
```

Let us try to calculate $f'(0)$ from first principles; we must take $f(0) = 0$, which is required to make f continuous.

```
In[23]:= Plot[f[x], {x, -0.5, 0.5}];
In[24]:= f[h]/h
                1
Out[24]= h Sin[-]
                h
In[25]:= Limit[%, h -> 0]
Limit::nlm: Could not find definite limit.
                       1
Out[25]= Limit[h Sin[-], h -> 0]
                       h
```

What we run up against is a limit that *Mathematica* cannot evaluate; thus it cannot find $f'(0)$. Of course, we know this last limit to be zero; thus $f'(0) = 0$. See Figure 3.2.

Example 5. Suppose we define $w(x) = [[x]]$, the greatest integer function. *Mathematica* cannot differentiate w, even though its derivative is 0 everywhere, except at each integer n where w is not even continuous, let alone differentiable:

```
In[26]:= w[x_] := Floor[x]
In[27]:= Plot[w[x], {x, -3, 3}];
In[28]:= w'[x]
Out[28]= w'[x]
```

Suppose we try to evaluate $w'(2.5)$ from first principles:

```
In[29]:= Limit[(w[2.5 + h] - w[2.5])/h, h -> 0]
Out[29]= Floor'[2.5]
```

Mathematica's frustrating answer has us going in circles! Of course, the functions in the four preceeding examples hardly require a computer to be differentiated. *Mathematica* is most useful when differentiating complicated functions built up from simple functions that *can* be differentiated. See Exercises 3, 4 and 5 below.

Example 6. *Mathematica* can differentiate sums and tables directly. This can be very useful, especially when working with power series. Try the following examples:

3.2. DIFFERENTIATION WITH MATHEMATICA

```
In[30]:= s[x_] := Sum[x^i, {i, 10}]
In[31]:= s'[x]
Out[31]= 1 + 2 x + 3 x^2 + 4 x^3 + 5 x^4 + 6 x^5 + 7 x^6 + 8 x^7 + 9 x^8 + 10 x^9
```

Even if no definite upper limit is given, *Mathematica* will still differentiate the general term:

```
In[32]:= t[x_] := Sum[x^i, {i, n}]
In[33]:= t'[x]
Out[33]= Sum[i x^(-1 + i), {i, n}]
```

Something similar happens for tables:

```
In[34]:= v[x_] := Table[Sin[i x], {i, 10}]
In[35]:= v'[x]
Out[35]= {Cos[x], 2 Cos[2 x], 3 Cos[3 x], 4 Cos[4 x], 5 Cos[5 x],
   6 Cos[6 x], 7 Cos[7 x], 8 Cos[8 x], 9 Cos[9 x], 10 Cos[10 x]}
```

However, if no definite upper limit is given for the iterator i, *Mathematica* returns no useful derivatives:

```
In[36]:= w[x_] := Table[Sin[i x], {i, n}]
In[37]:= w'[x]
General::iterbounds: Iterator {i, n} does not have appropriate bounds.
General::iterbounds: Iterator {i, n} does not have appropriate bounds.
General::iterbounds: Iterator {i, n} does not have appropriate bounds.
General::stop: Further output of General::iterbounds
    will be suppressed during this calculation.
Out[37]= Table[D[Sin[i x], x], {i, n}]
```

Example 7. Another type of differentiation that *Mathematica* can do is implicit differentiation. Consider for example the equation of a circle

$$x^2 + y^2 = 9.$$

Differentiating this equation implicitly results in a formula for dy/dx in terms of both x and y. *Mathematica* can do this as well, but you *must indicate that y is to be considered as a function of x*. To do this, simply enter y[x] for y:

```
In[38]:= D[x^2 + y[x]^2 == 9, x]
Out[38]= 2 x + 2 y[x] y'[x] == 0
```

Then you can simply solve this equation for y':

```
In[39]:= Solve[%, y'[x]]
                    x
Out[39]= {{y'[x] -> -(----)}}
                   y[x]
```

You must remember to indicate whether or not a variable is to be considered as a function of another variable, or *Mathematica* will simply assume it is a constant and calculate its derivative to be 0. Compare:

```
In[40]:= D[f[x], t]
Out[40]= 0
In[41]:= D[f[x[t]], t]
Out[41]= f'[x[t]] x'[t]
```

Example 8. In general, finding higher order derivatives can become quite messy. This is precisely where a computer becomes most helpful. Consider the function f defined by
$$f(x) = \frac{2x}{x^2+1}.$$
It is a simple matter to calculate f' and f'' with *Mathematica*:

```
In[42]:= f[x_] := (2 x)/(1 + x^2)
In[43]:= f'[x]
              2
          -4 x           2
Out[43]= --------- + ------
                2 2        2
          (1 + x )    1 + x
In[44]:= f''[x]
              3
          16 x           12 x
Out[44]= --------- - ---------
                2 3         2 2
          (1 + x )    (1 + x )
```

We could proceed in this fashion, by using repeatedly more primes to obtain higher derivatives; but this is unnecessary since *Mathematica* offers the function `Derivative` which can be used to calculate any order derivative :

```
In[45]:= ?Derivative
f' represents the derivative of a function f of one argument.
```

3.2. DIFFERENTIATION WITH MATHEMATICA

`Derivative[n1,n2, ...][f]` is the general form, representing a function obtained from f by differentiating n1 times with respect to the first argument, n2 times with respect to the second argument, and so on.

Note that `Derivative` can also be used to calculate partial derivatives. However, we shall use it mostly to calculate higher order derivatives of a function of one variable. For example the fifth derivative of f:

```
In[45] := Derivative[5][f][x]
```

$$\text{Out[45]} = \frac{-7680\, x^6}{(1+x)^{26}} + \frac{11520\, x^4}{(1+x)^{25}} - \frac{4320\, x^2}{(1+x)^{24}} + \frac{240}{(1+x)^{23}}$$

```
In[46] := Simplify[%]
```

$$\text{Out[46]} = \frac{-240\, (-1+x)\, (1+x)\, (1-4x+x^2)\, (1+4x+x^2)}{(1+x)^{26}}$$

MATHEMATICA FUNCTIONS INTRODUCED IN THIS SECTION
Derivative

Exercises:

1. What is
$$\lim_{h \to 0} \frac{f(a-h) - f(a)}{h}?$$
Does *Mathematica* give the right answer?

2. What is
$$\lim_{h \to 0} \frac{f(a+h) + f(a)}{\sqrt{h^2}}?$$
Does *Mathematica* give the right answer? (Be careful!)

3. Use *Mathematica* to find derivatives of the following functions:

   (a) $f(x) = e^{e^{e^{e^x}}}$

   (b) $f(x) = u(x)v(x)w(x)$ (triple product rule)

   (c) $f(x) = \dfrac{\sin\sqrt{x^2 + \frac{1}{x^2}} \log(x^2 + x^4)}{x + e^x - \sqrt{x}}$

(d) $f(x) = \dfrac{x^{15} + x^7 - x^3 + 1}{x^{17} + x^9 - x^4 + 3}$

4. For f as in Exercise 3 c), plot f' on $[0.1, 2]$ to determine how many critical points f has in $[0.1, 2]$.

5. For f as in Exercise 3 d), plot f' on $[-3, 3]$ to determine how many critical points f has in $[-3, 3]$. What kind of discontinuity does f have at $x = -1$?

6. With s, t and v as in Example 6, use *Mathematica* to find the *second* derivatives of s, t and v.

7. Differentiate implicity to find the indicated derivative at the indicated point:

    (a) $x + xy + y = -3$; y' at $(x, y) = (1, -2)$
    (b) $x + xy + y = -3$; y'' at $(x, y) = (1, -2)$
    (c) $x^3 - 3xy + y^3 = -1$; y' at $(x, y) = (2, -3)$
    (d) $x^3 - 3xy + y^3 = -1$; y'' at $(x, y) = (2, -3)$

8. Use `Derivative` to find the indicated derivative for each of the following functions:

    (a) $f(x) = \dfrac{2x}{x^2 + 1}$; $f^{(5)}(3)$
    (b) $f(x) = \dfrac{x^{11} + x^6 - 2}{x^7 + x^3 + 3}$; $f^{(4)}(-1)$
    (c) $f(x) = x^2 e^x$; $f^{(5)}(x)$
    (d) Derive a general formula for the *n*th derivatie of $f(x) = x^2 e^x$.

3.3 Tangents and Normals

This section consists of many examples covering typical problems that involve tangents and normals. We shall make use of the defined functions in `Chap3.m` to simplify our calculations and to sketch graphs whenever possible. First we load the package for Chapter 3.

`In[1]:= << Chap3.m`

We shall make extensive use of the functions `tangent` and `normal`. We can remind ourselves of their syntax by querying *Mathematica*:

3.3. TANGENTS AND NORMALS

```
In[2] := ?tangent
tangent
tangent/: tangent[f_, a_][x_] := f'[a] (x - a) + f[a]

In[2] := ?normal
normal
                                    a - x
normal/: normal[f_, a_][x_]  :=    -----  + f[a]
                                    f'[a]
```

Note that in the definition of **normal** no special consideration is given for the case $f'(a) = 0$, where, of course, the normal is a vertical line and the formula for **normal** no longer makes sense. We must be wary of any calculations that involve this case!

Example 1. Find the tangents to the hyperbola $xy = 1$ passing through the point $(-1, 1)$. This point is *off* the hyperbola; we expect *two* tangent lines from the hyperbola to pass through this point. See Figure 3.3. We work as follows:

```
In[2] := f[x_] := 1/x
In[3] := Solve[tangent[f,a][-1]==1,a]   (* (-1,1) is on the tangent *)

              2 + 2 Sqrt[2]         2 - 2 Sqrt[2]
Out[3]= {{a -> -------------}, {a -> -------------}}
                    2                      2
In[4] := tangent[f, a][x] /. %

                           -(2 + 2 Sqrt[2])
                        4 (---------------- + x)
                2                2
Out[4]= {-------------  -  -------------------------,
          2 + 2 Sqrt[2]                 2
                              (2 + 2 Sqrt[2])

                           -(2 - 2 Sqrt[2])
                        4 (---------------- + x)
                2                2
>        -------------  -  -------------------------}
          2 - 2 Sqrt[2]                 2
                              (2 - 2 Sqrt[2])

In[5] := Simplify[%]

          2 + 2 Sqrt[2] - x    -2 + 2 Sqrt[2] + x
Out[5]= {------------------,  ------------------}
```

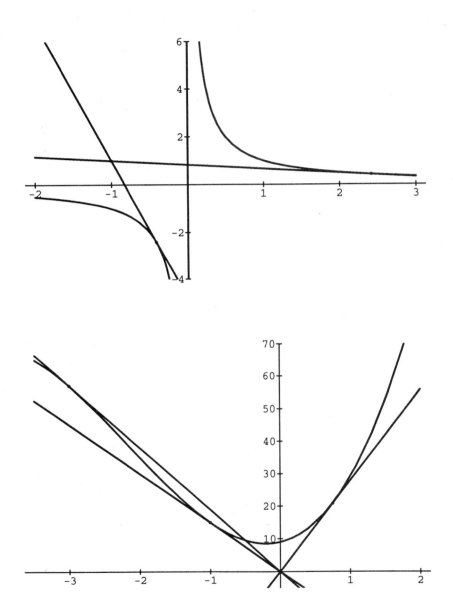

Figure 3.3: Top: there are two tangents to the hyperbola $xy = 1$ passing through the point $(-1, 1)$. Bottom: there are three tangents to the cubic $y = 2x^3 + 13x^2 + 5x + 9$ passing through the point $(0, 0)$.

3.3. TANGENTS AND NORMALS

```
                3 + 2 Sqrt[2]           -3 + 2 Sqrt[2]

In[6]:= Plot[{%[[1]], %[[2]], f[x]}, {x, -2, 3},
             PlotRange -> {-4, 6}];
```

This last output should resemble Figure 3.3.

Example 2. Find all the tangents to the graph of
$$y = 2x^3 + 13x^2 + 5x + 9$$
passing through the origin. The point $(0,0)$ is not on the cubic; there could be as many as three tangents to the cubic passing through $(0,0)$. See Figure 3.3.

```
In[7]:= f[x_] := 2 x^3 + 13 x^2 + 5 x + 9
In[8]:= Solve[tangent[f, a][0] == 0, a]   (* (0, 0) is on the tangent *)
                                3
Out[8]= {{a -> -1}, {a -> -3}, {a -> -}}
                                4
In[9]:= tangent[f, a][x] /. %
                                                3
                                      223 (-(-) + x)
                              669             4
Out[9]= {15 - 15 (1 + x), 57 - 19 (3 + x), --- + ---------------}
                                      32             8
In[10]:= Plot[{%[[1]], %[[2]], %[[3]], f[x]}, {x, -3.5, 2},
              PlotRange -> {-5, 70}];
```

This final output should resemble the bottom graph of Figure 3.3.

Example 3. Let $f(x) = (\sqrt{k} - \sqrt{x})^2$. Prove that the sum of the intercepts on the co-ordinate axes of any tangent to f is a constant.

```
In[11]:= f[x_] := (Sqrt[k] - Sqrt[x])^2
In[12]:= tangent[f, a][x]
                              2   (-Sqrt[a] + Sqrt[k]) (-a + x)
Out[12]= (-Sqrt[a] + Sqrt[k])  - -------------------------------
                                             Sqrt[a]
In[13]:= tangent[f, a][0]    (* y intercept *)
                                                            2
Out[13]= Sqrt[a] (-Sqrt[a] + Sqrt[k]) + (-Sqrt[a] + Sqrt[k])
In[14]:= Solve[tangent[f, a][x] == 0, x]   (* x intercept *)
Out[14]= {{x -> Sqrt[a] Sqrt[k]}}

In[15]:= Out[13] + Sqrt[a] Sqrt[k]    (* sum of intercepts *)
```

```
Out[15]= Sqrt[a] (-Sqrt[a] + Sqrt[k]) + (-Sqrt[a] + Sqrt[k])2 +

>     Sqrt[a] Sqrt[k]
In[16]:= Simplify[%]
Out[16]= k
```

Thus the sum of the intercepts of any tangent to f is a constant, k. See Figure 3.4.
Example 4. Find the tangent and normal lines to the parabola with equation $y = x^2$ at the point $(-1, 1)$. We can use the defined functions `normalplot` and `tangentplot`. Note that `normalplot` sets the aspect ratio at automatic, which is necessary if you wish to produce graphs with normals that actually look perpendicular to the curve.

```
In[17]:= ?normalplot
normalplot
normalplot/: normalplot[f_, a_, {b_, c_}] :=

>     Plot[{f[x], normal[f, a][x]}, {x, b, c}, AspectRatio -> Automatic]

In[17]:= f[x_] := x^2
In[18]:= normalplot[f, -1, {-2, 2}];
In[19]:= tangentplot[f, -1, {-2, 2}];
In[20]:= Show[%%, %];
```

Your graph should resemble the bottom picture of Figure 3.4. To obtain the actual equations of these tangent and normal lines we can use `tangent` and `normal`, and simlify the results:

```
In[22]:= tangent[f, -1][x]
Out[22]= 1 - 2 (1 + x)
In[23]:= normal[f, -1][x]
              -1 - x
Out[23]= 1 -  ------
                2
In[24]:= Simplify[{tangent[f, -1][x], normal[f, -1][x]}]
                          3 + x
Out[24]= {-1 - 2 x, -----}
                            2
```

So far all of these examples have been very routine. The remaining examples will be slightly more complicated, using more of the *Mathematica*'s capabilities.

Example 5. One thing we can do is graph *lots* of tangents. There is a defined function, `tangentfamily`, in `Chap3.m` to do this:

3.3. TANGENTS AND NORMALS

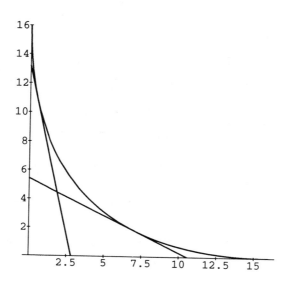

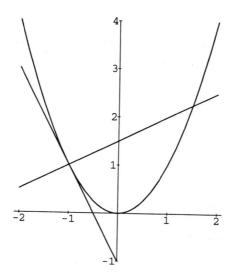

Figure 3.4: Top: the sum of the intercepts of any tangent to the graph of $\sqrt{x}+\sqrt{y}=\sqrt{k}$ is a constant. Bottom: tangent and normal to $y=x^2$ at the point $(-1,1)$.

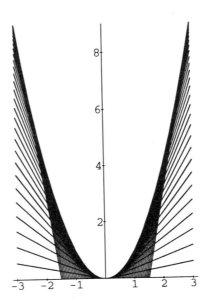

Figure 3.5: The family of tangents to the parabola with equation $y = x^2$; the curve is said to be the *envelope* of its tangents.

```
In[25]:= ?tangentfamily
tangentfamily[f, {a, b, stepsize}] plots f and its tangents on the
interval {a, b}, plotting one tangent per stepsize in the interval
{a, b}. The default stepsize is 0.1.
```

We will illustrate it on the parabola with equation $y = x^2$.

```
In[25]:= tangentfamily[f, {-3, 3}];
In[26]:= Show[%, AspectRatio -> Automatic, PlotRange -> {0, 9}];
```

Note that the graph looks better if the aspect ratio is set at automatic — this spreads the tangents out more. See Figure 3.5. By looking at your graphs, determine which points in the plane are *not* on any tangent of $y = x^2$. We say that the function f is the *envelope* of its tangents. Indeed, looking at Figure 3.5, you can hardly see the graph for all the tangent lines; but the shape of the graph is highlighted by the tangents.

Example 6. In addition to `tangentfamily`, there is a similar defined function showing normals to a curve, called `normalfamily`:

```
In[27]:= ?normalfamily
```

3.3. TANGENTS AND NORMALS

normalfamily[f, {a, b, stepsize}, {c, d}] plots f and its normals on the interval {a, b}, with PlotRange set at {c, d}. One normal is plotted per stepsize in the interval {a, b}; the default stepsize is 0.1.

We can illustrate this function first on the parabola $y = x^2$:

```
In[27]:= normalfamily[f, {-6, 6}, {-2, 10}];   (* f as before *)
```

Your output should look something like the top of Figure 3.6. Note the interesting pattern the normals create: they highlight a curve which is the envelope of all the normals. Let us plot the normals to $y = \sin x$ to see if an interesting pattern emerges again:

```
In[28]:= normalfamily[Sin, {-6, 6}, {-6, 6}];
```

Your output should look something like the bottom of Figure 3.6.

Example 7. Two curves are said to be *orthogonal*, Greek for perpendicular, if the angle between their tangents at all intersection points is a right angle. This means the tangent of one curve is the normal of the other, and vice-versa, at the intersection. To illustrate such a situation, consider the ellipse with equation

$$4x^2 + 9y^2 = 45$$

and the hyperbola with equation

$$x^2 - 4y^2 = 5.$$

We can use a parametric plot to graph the ellipse, but we will have to plot the hyperbola by pieces:[1]

```
In[29]:= ParametricPlot[{Sqrt[45] Cos[t]/2, Sqrt[5] Sin[t]},{t,0,2Pi},
           AspectRatio -> Automatic];
In[30]:= Plot[{Sqrt[x^2 - 5]/2, -Sqrt[x^2 - 5]/2}, {x, Sqrt[5], 8}];
In[31]:= Plot[{Sqrt[x^2 - 5]/2, -Sqrt[x^2 - 5]/2}, {x, -8, -Sqrt[5]}];
In[32]:= Show[Out[29], %, %%];
```

As you can see, there are four intersection points, which we can readily find with *Mathematica*:

---

[1] If you are familiar with *hyperbolic* trigonometric functions, then you could use them to plot the hyperbola parametrically.

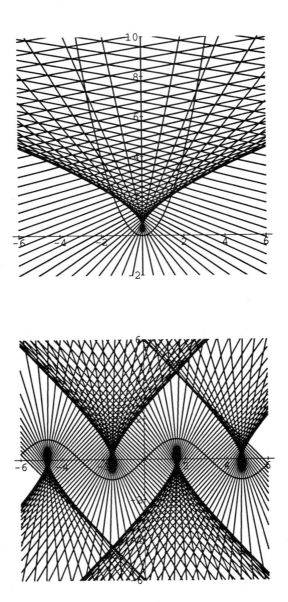

Figure 3.6: Families of normals to $y = x^2$ (top) and $y = \sin x$ (bottom). Note the interesting curves which form the envelopes of the normals.

3.3. TANGENTS AND NORMALS

```
In[33] := Solve[{4 x^2 + 9 y^2 == 45, x^2 - 4 y^2 == 5}, {x, y}]
Out[33]= {{x -> 3, y -> 1}, {x -> -3, y -> 1}, {x -> 3, y -> -1},

>       {x -> -3, y -> -1}}
In[34] := ListPlot[{{3, 1}, {-3, 1}, {3, -1}, {-3, -1}}];
In[35] := Show[Out[32], %];
```

So far we have only constructed the graphs and highlighted the points of intersection. To check if the curves are actually orthogonal at these intersection points, let us check if the tangent to one at $(x, y) = (3, -1)$, say, is the normal to the other at the same point. (To do the calculations, functions f, g, h and k are defined first.)

```
In[36] := f[x_] := Sqrt[45 - 4 x^2]/3;  g[x_] := - Sqrt[45 - 4 x^2]/3
In[37] := h[x_] := Sqrt[x^2 - 5]/2;  k[x_] := - Sqrt[x^2 - 5]/2
In[38] := tangent[f, 3][x]
                4 (-3 + x)
Out[38]= 1 -  ------------
                    3
In[39] := normal[h, 3][x]
                4 (3 - x)
Out[39]= 1 +  -----------
                    3
```

Similar calculations at the other three points of intersection lead to similar results. The two curves, with normals and tangents at the four intersection points are pictured in Figure 3.7.

Example 8. Minimum distance in the plane is measured perpendicularly. Thus we can use normals to find the point on the parabola with equation $y = x^2/8$ closest to the point $(2, 4)$, which is not on the parabola. First we can plot the curve and the point to see what the situation is. See Figure 3.8, top part.

```
In[40] := f[x_] := x^2/8
In[41] := Plot[f[x], {x, -1, 7}];
In[42] := ListPlot[{{2, 4}}];
In[43] := Show[%, %%];
```

We can find the point $x = a$ for which the normal to f passes through $(2, 4)$ as follows:

```
In[44] := Solve[normal[f, a][2] == 4, a]
                            (2 I)/3 Pi              (4 I)/3 Pi
Out[44]= {{a -> 4}, {a -> 4 E          }, {a -> 4 E          }}
In[45] := normalplot[f, 4, {-1, 7}];   (* only the real root is relevant *)
In[46] := Show[%, Out[43]];
```

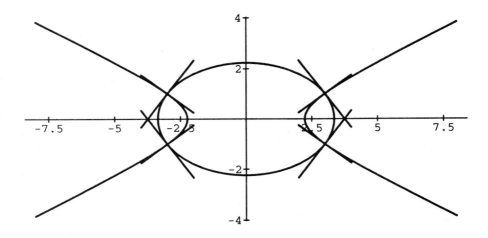

Figure 3.7: The ellipse $4x^2 + 9y^2 = 45$ and the hyperbola $x^2 - 4y^2 = 5$ are orthogonal curves.

Thus the point on f closest to the point $(2,4)$ is the point $(4,2)$. See Figure 3.8

Example 9. In a fashion similar to the calculations of the previous example, we can find the closest pair of points, one from each curve, on two different curves. Consider the two parabolas f and g with equations

$$f(x) = x^2 + 2x + 2 \text{ and } g(x) = -x^2 + 6x - 4.$$

```
In[47]:= f[x_] := x^2 + 2 x + 2; g[x_] := -x^2 + 6 x - 4
In[48]:= Plot[{f[x], g[x]}, {x, -2, 4}];
```

Your output should look something like the bottom part of Figure 3.8, without the line. Suppose the points $(a, f(a))$ and $(b, g(b))$ are the two points closest together. The first point must be on the normal to g at $x = b$; the second point must be on the normal to f at $x = a$. This gives us a system of two equations to solve for a and b:

```
In[49]:= Solve[{normal[f, a][b] == g[b], normal[g, b][a] == f[a]}, {a, b}]

Out[49]=  (*  ... pages and pages of output scroll by ... *)
```

3.3. TANGENTS AND NORMALS

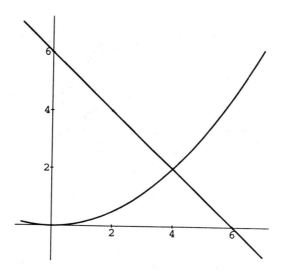

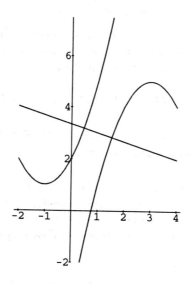

Figure 3.8: Shortest distance, from a point to a curve (top), and between curves (bottom), is measured perpendicularly.

```
In[50]:= % // N
Out[50]= {{a -> 1. + 1.41421 I, b -> 1. + 1.41421 I},

>       {a -> 1. - 1.41421 I, b -> 1. - 1.41421 I},

>       {a -> 0.475687, b -> 1.52431},

>       {a -> -1.73784 - 0.365018 I, b -> 3.73784 + 0.365018 I},

>       {a -> -1.73784 + 0.365018 I, b -> 3.73784 - 0.365018 I}}
```

Only one of these solutions is real. The points on f and g, closest together, are approximately:

```
In[51]:= {{a, f[a]}, {b, g[b]}} /. %[[3]]
Out[51]= {{0.475687, 3.17765}, {1.52431, 2.82235}}
```

Exercises:

1. At what points on the function f defined by $f(x) = x^3 + 5$ is its tangent

    (a) parallel to the line with equation $12x - y = 17$?

    (b) perpendicular to the line with equation $x + 3y = 2$?

    (c) parallel to the x−axis?

2. Find the equations of the tangents to the parabola with equation $y = -x^2 + 4x$ passing through the point $(2, 40)$.

3. Find the equations of the tangents to the quartic with equation $y = x^4 - 3x^2 + x$ passing through the point $(-3, 2)$. (You will have to approximate your answers.) Plot a graph of y, showing the points of contact and the tangents, and the point $(-3, 2)$.

4. Consider the curve with equation
$$x^{2/3} + y^{2/3} = a^{2/3}.$$

    (a) Define a function f which represents the top half of the curve.

    (b) Show that the sum of the squares of the intercepts on the co−ordinate axes of any tangent to f is a constant.

5. Look over the family of normals to the parabola $y = x^2$, Figure 3.6. On how many normals to the parabola does a point lie if

    (a) the point is *below* the envelope to the normals?

3.3. TANGENTS AND NORMALS

(b) the point is *on* the envelope to the normals?

(c) the point is *above* the envelope to the normals?

6. Find the point(s) on the parabola $y = x^2$ closest to the following points:

   (a) $(3, -1)$

   (b) $(8, 7)$

   (c) $(-32, 25/2)$

7. Consider the parabola with equation $y = x^2$ and an arbitrary point $(0, b)$ on the y−axis. For which values of b will there be

   (a) exactly one point on the parabola closest to $(0, b)$?

   (b) exactly two points on the parabola closest to $(0, b)$?

   (c) exactly three points on the parabola closest to $(0, b)$?

   Recall: the y−axis *is a normal* to the parabola at the vertex $(0, 0)$.

8. Use **normalfamily** to plot the family of normals to the following functions for the specified plot range:

   (a) e^x; $-5 \leq x \leq 3$, $0 \leq y \leq 8$

   (b) $\arctan x$; $-8 \leq x \leq 8$, $-8 \leq y \leq 8$

   (c) $3\sqrt{1 - x^2/16}$; $-3.9 \leq x \leq 3.9$, $-4 \leq y \leq 4$

   (d) Based on the graphical output from parts a), b), and c), determine in each case, the points in the plane which are:

      i. on exactly one normal to the given curve

      ii. on exactly two normals

      iii. on three or more normals.

9. Find the shortest distance between the graphs of the two functions
$$f(x) = -10/x \text{ and } g(x) = x^3 - 3x.$$

   Plot graphs first to judge where the two points closest together might be. Warning: if you are solving this question by using the method of Example 9, be sure to approximate the solutions — there will be 15 of them!

10. Show that the ellipse and hyperbola with equations
$$2x^2 + y^2 = 20 \text{ and } 4y^2 - x^2 = 8,$$

    respectively, are orthogonal. (Hint: by symmetry it is only necessary to consider points in the first quadrant.)

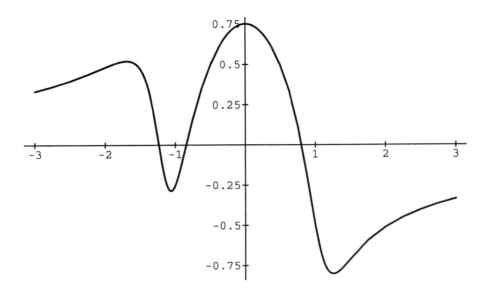

Figure 3.9: Graphing a continuous function on a closed interval to see where its extreme values are.

3.4 Applications of the Derivative

One of the most important applications of calculus is finding extreme values — maximum or minimum – of a function. If the function is continuous and restricted to a closed interval, then extreme values must always exist. In this case, one way to see where they are is to plot the function on the given interval. *Mathematica* can do this very well. However, analysis is still required if you wish to find the exact co–ordinates of the extreme points.

Example 1. Consider the function f defined by

$$f(x) = \frac{3 - 4x^2 - x^6 - x^9}{4 + x^7 + x^{10}}.$$

If we plot its graph on the interval $[-3, 3]$, we can see immediately that it has two relative maxima and two relative minima. See Figure 3.9.

```
In[1]:= f[x_] := (3 - 4 x^2 - x^6 - x^9)/(4 + x^7 + x^(10))
In[2]:= Plot[f[x], {x, -3, 3}];
```

To find the values of the critical points, where the derivative is zeo or fails to exist, we can do the usual analysis:

3.4. APPLICATIONS OF THE DERIVATIVE

```
In[3]:= Solve[f'[x] == 0, x] // N
Out[3]= {{x -> 0.}, {x -> -1.69323}, {x -> -1.06201},
>       {x -> -0.921237 - 1.23407 I}, {x -> -0.921237 + 1.23407 I},
>       {x -> -0.875939 - 0.600961 I}, {x -> -0.875939 + 0.600961 I},
>       {x -> -0.38643 - 0.835818 I}, {x -> -0.38643 + 0.835818 I},
>       {x -> 0.369626 - 0.906101 I}, {x -> 0.369626 + 0.906101 I},
>       {x -> 0.443134 - 1.70658 I}, {x -> 0.443134 + 1.70658 I},
>       {x -> 0.729244 - 0.484287 I}, {x -> 0.729244 + 0.484287 I},
>       {x -> 1.2562}, {x -> 1.39112 - 0.895054 I},
>       {x -> 1.39112 + 0.895054 I}}
```

We can evaluate f at the four real critical points to determine its extreme values:

```
In[4]:= f[x] /. {%[[1]], %[[2]], %[[3]], %[[16]]}
Out[4]= {0.75, 0.521957, -0.285411, -0.802886}
```

Thus the maximum value of f on $[-3, 3]$ is 0.75; and the minimum value of f on $[-3, 3]$ is -0.802886.

Example 2. As an example of a word problem involving extreme values, consider question 22, on p 122 of Edwards and Penney[3]:

> A wire of length 100 cm is to be cut into two pieces. One piece is bent into a circle, the other into a square. How should the cut be made to maximize the sum of the areas of the square and the circle? How should it be done to minimize that sum?

Mathematica will not help you set this problem up, but once you do have it set up then you can use *Mathematica* to analyse the overall situation much more thoroughly than with only paper and pencil. First of all, let the cut be made at x, $0 \leq x \leq 100$:

Suppose the piece on the left of length x is bent into a circle, and the piece on the right of length $100 - x$ is bent into a square. Let $a(x)$ be the sum of the circle's and square's area. Then

$$a(x) = \frac{x^2}{4\pi} + \frac{(100 - x)^2}{16}.$$

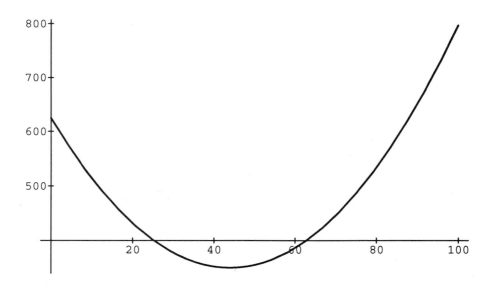

Figure 3.10: The minimum of $a(x)$ is at the critical point; the maximum of $a(x)$ is at the right end point.

We can plot a to get an immediate overall picture of the problem. See Figure 3.10.

```
In[5]:= a[x_] := x^2/(4 Pi) + (100 - x)^2/16
In[6]:= Plot[a[x], {x, 0, 100}]
Out[6]= -Graphics-
```

You can see that the maximum total area is at the right end point $x = 100$ (corresponding to bending the whole wire into a circle), and the minimum total area is at the critical point. We can find the critical point by setting $a'(x) = 0$:

```
In[7]:= Solve[a'[x] == 0, x]
                100 Pi
Out[7]= {{x -> ------}}
                4 + Pi
In[8]:= % // N
Out[8]= {{x -> 43.9901}}
```

Thus the minimum total area occurs with a cut at about $x = 43.99$.

With a computer we can do more. For instance, a defined function, **wire**, is put into the package **Chap3.m** so that you can actually see the effect of cutting the wire in different places. Use **wire** for different values of x, say $x = 30, 45, 70, 90$:

3.4. APPLICATIONS OF THE DERIVATIVE

```
In[9]  := wire[30];
In[10] := wire[45];
In[11] := wire[70];
In[12] := wire[90];
```

These four cases are illustrated in Figures 3.11 and 3.12 for your reference. Once you have defined a function such as `wire` which captures the problem for any value of the parameter x, then it is an easy matter to animate a series of pictures as in Figures 3.11 and 3.12 to form a dynamic model of the problem. You can create your own animation in DOS as follows:

$$\text{Animate[Table[wire[i], \{i, 0, 100, 5\}]]}.$$

This animation is on file as *wire.ani*. For NeXT workstations we just string the individual images together and hide all but the first one before activating.

Example 3. Since *Mathematica* can handle implicit differentiation it is possible to do related rates problems directly on computer. For example, consider the typical problem, question 45 on p 143 of Edwards and Penney[3]:

> A ladder 41 ft long has been leaning against a vertical wall. It begins to slip, so that its top slides down the wall while its bottom moves along the ground; the bottom moves at a constant speed of 10 ft/s. How fast is the top of the ladder moving when it is 9 ft above the ground?

See Figure 3.13. We can solve this problem with *Mathematica* as follows:

```
In[13] := D[x[t]^2 + y[t]^2 == 41^2, t]
Out[13]= 2 x[t] x'[t] + 2 y[t] y'[t] == 0
In[14] := Solve[x^2 + 9^2 == 41^2, x]
Out[14]= {{x -> 40}, {x -> -40}}
In[15] := %% /. {x[t] -> 40, y[t] -> 9, x'[t] -> 10}
Out[15]= 800 + 18 y'[t] == 0
In[16] := Solve[%, y'[t]]
                  400
Out[16]= {{y'[t] -> -(---)}}
                   9
```

Thus the ladder will be moving down the wall at 44.4 ft/s when it is 9 ft above the ground. As with Example 2, it is possible to define a function — in this case it is called `ladder` — which captures the problem at any time, t, measured in seconds, with $t = 0$ corresponding to the vertical position of the ladder. You can find `ladder` in the package `Chap3.m`. Figure 3.13 was made by using `ladder[3.4]`:

```
In[17] := ladder[3.4];     (* t = 3.4 s; x = 34 ft *)
```

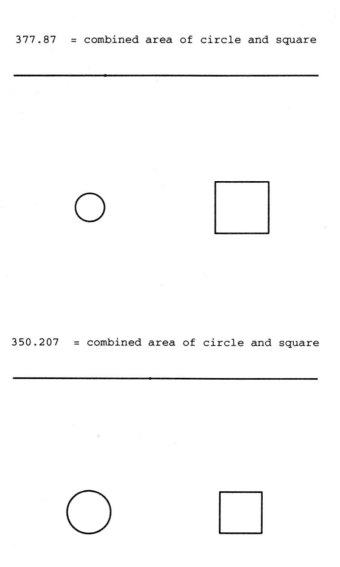

Figure 3.11: Circle and square produced by cutting the wire at $x = 30$ (top) and $x = 45$ (bottom).

3.4. APPLICATIONS OF THE DERIVATIVE

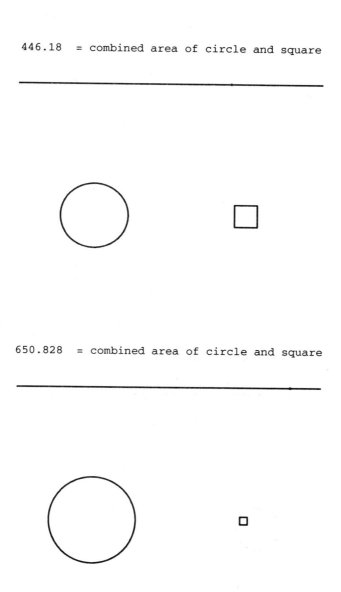

Figure 3.12: Circle and square produced by cutting the wire at $x = 70$ (top) and $x = 90$ (bottom). The maxmimum enclosed area is when the entire wire is bent into a circle.

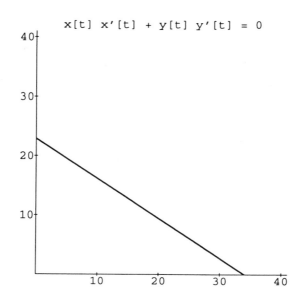

Figure 3.13: As the ladder moves, the rates at which y and x change are related.

It is also possible to animate a table of ladder[i] to obtain a visual representation of the problem. There is an animation file to show this in DOS; it is *ladder.ani*.

FUNCTIONS CONTAINED IN THE PACKAGE Chap3.m		
tangent	normal	tangentplot
normalplot	tangentfamily	normalfamily
secant	wire	ladder

Exercises:

1. Plot each of the following functions, f, on the indicated intervals, $[a, b]$, to look for extreme valules. By finding the critical points if necessary, determine the maximum and minimum values of f on $[a, b]$.

    (a) $f(x) = |x|$ on $[-2, 4]$
    (b) $f(x) = x^4 - x^3 - 3x^2 + 7$, on $[-2, 3]$
    (c) $f(x) = \dfrac{x^5 - x^3 + 5x^2 - 6}{x^2 + 4}$, on $[-2, 2]$
    (d) $f(x) = |x^5 - 4x^4 + 6x^3 - x + 2|$, on $[-1, 1]$
    (e) $f(x) = x^{7/3} - 7x^{4/3} + 15x^{1/3}$, on $[-1, 4]$

3.4. APPLICATIONS OF THE DERIVATIVE

2. Consider question 38, p 123, of Edwards and Penney[3]:

    Two vertical poles are each 10 ft tall and stand 10 ft apart. Find the length of the shortest rope that can reach from the top of one pole to a point between them on the ground and then on to the other pole.

    This is a nice problem, but for the archaic units which we hope engineers will eventually stop using!

    (a) Set up the problem by placing one pole at $(0, 0)$ and the other at $(10, 0)$. Suppose the rope goes from the top of one pole at $(0, 10)$ down to a point $(x, 0)$ and then to the top of the other pole at $(10, 10)$. What is the length of the rope, $l(x)$?

    (b) Plot $l(x)$ on the interval $[0, 10]$. Which value of x minimizes l?

    (c) Now suppose the pole at $(0, 0)$ is actually k ft tall. What is $l(x)$ now? Which value of x minimizes l now? Your answer will depend on k.

    (d) Plot the critical value found in part c) against k, for $0 \leq k \leq 50$. What happens to the point on the ground where the rope touches as $k \to \infty$?

    (e) Define a linear function joining the points $(0, -k)$ and $(10, 0)$. What is the x–intercept of this line? Why can this linear function be used to solve the original problem?

    (f) Now suppose the two vertical poles both have undetermined height, say one is k ft tall, and the other is m ft tall, and that they are an unspecified distance a ft apart. Find the length of the shortest rope that can reach from the top of one pole to a point between them on the ground and then to the top of the other pole.

3. Consider question 27, p 123, of Edwards and Penney[3]:

    A printing company has eight presses, each of which can print 3600 copies per hour. It costs $5.00 to set up each press for a run, and $10 + 6n$ dollars to run n presses for one hour. How many presses should be used in order to print 50,000 copies of a poster most profitably?

    (a) Solve this problem in the usual way by finding the cost function, $c(n)$, of using n presses to print the 50,000 copies. Plot c for $1 \leq n \leq 8$. At which value of n is c minimized?

    (b) Now consider the general case of having to print a copies. What is $c(n)$ in this case? Plot $c(1), \ldots, c(8)$ for a ranging from 0 to 100,000.

    (c) What is the minimum number of copies required to make the use of 2 presses economical? What is the minimum number of copies required to make the use of 3 presses economical?

(d) Give a complete breakdown on the most economical number of presses to be used to print a copies.

(e) What is the mimimum number of copies required before it becomes economical to use all 8 presses?

4. Similar to Example 3, set up a function in terms of one parameter that describes the situation of question 53, p 143, of Edwards and Penney[3]. Plot different states for different values of the parameter. Can you solve the problem just by looking at your plots?

Chapter 4

Methods of Approximation

> *No barber shaves so close but another finds work.* —G.Herbert, Outlandish Proverbs, 1640

We return to the concept that introduced the derivative of a function at the beginning of Chapter 3, as the best linear approximation to the function.

4.1 Linear Approximation

If the function f is differentiable at $x = a$, then the equation of the tangent line to f at $x = a$ is

$$y = f'(a)(x - a) + f(a).$$

For values of x "close" to a, the tangent line and the graph of $y = f(x)$ are "close" together. See Figure 4.1. Therefore the tangent line can be used to approximate values of $f(x)$ if x is near a:

$$f(x) \approx f'(a)(x - a) + f(a).$$

Such approximations are called *linear approximations*. In the package for Chapter 4 we have included functions to calculate linear approximations and to graph tangent lines.

```
In[1]:= Needs["Tutorial`Chap4`"]
In[2]:= ?tangent*
tangent     tangentplot
In[2]:= ?tangent
tangent
tangent/: tangent[f_, a_][x_] := f'[a] (x - a) + f[a]
```

103

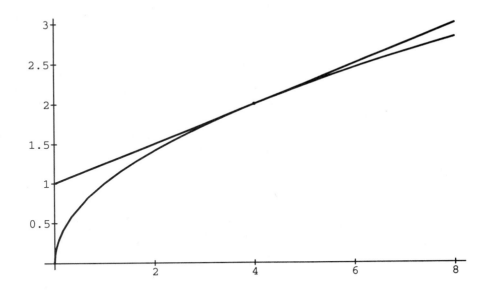

Figure 4.1: The tangent line to $y = \sqrt{x}$ at $x = 4$.

```
In[2]:= ?tangentplot
tangentplot
tangentplot/: tangentplot[f_, a_, {b_, c_}] :=

>       Plot[{f[x], tangent[f, a][x]}, {x, b, c}]
```

Example 1. Suppose $f(x) = \sqrt{x}$. What is the tangent to f at $x = 4$?

```
In[2]:= tangent[Sqrt, 4][x]
              -4 + x
Out[2]= 2 +  ------
                4
In[3]:= Simplify[%]
             x
Out[3]= 1 + -
             4
```

You can plot this line, and the function f, by using **tangentplot**. The output is Figure 4.1.

```
In[4]:= tangentplot[Sqrt, 4, {0, 8}];
```

4.1. LINEAR APPROXIMATION

Example 2. The defined function `tangent` is listable in x, so more than one linear approximation can be calculated in one input. Using the tangent line of Example 1, let us obtain approximations to

$$\sqrt{7.2}, \sqrt{2.5}, \sqrt{4.8}, \text{ and } \sqrt{3.95}.$$

If you then compare these approximations with the actual values you can see that the linear approximations are more accurate for values of x closer to 4.

```
In[5]:= tangent[Sqrt, 4][{7.2, 2.5, 4.8, 3.95}] // N
Out[5]= {2.8, 1.625, 2.2, 1.9875}
In[6]:= Sqrt[{7.2, 2.5, 4.8, 3.95}] // N
Out[6]= {2.68328, 1.58114, 2.19089, 1.98746}
```

The defined function `errorplot` of `Chap4.m` allows you to see the absolute error of the tangent line approximation to a function on a given interval. See Figure 4.2 and Figure 4.3.

```
In[7]:= errorplot[Sqrt, 4, {0, 8}];
In[8]:= errorplot[Sqrt, 4, {2, 6}];
In[9]:= errorplot[Sqrt, 4, {3, 5}];
In[10]:= errorplot[Sqrt, 4, {3.9, 4.1}];
```

Observe that on the very large interval, $[0, 8]$, the absolute error of the linear approximations can be as large as 1; but that on the small interval, $[3.9, 4.1]$, the absolute error of the linear approximations is no larger than 0.0002. Of course, the actual degree of accuracy depends not only on the interval, but also on the function f.

Exercises:

1. Find linear approximations for the following numbers:
   (a) $\sqrt[3]{25}$
   (b) $\sqrt{102}$
   (c) $(15)^{1/4}$
   (d) $(65)^{-2/3}$

2. What is the error for each of the approximations in Exercise 1?

3. Prove the following linear approximation formula for x near zero:

$$(1+x)^k \approx 1 + kx.$$

Use this formula to obtain a linear approximation to $(1.048)^{1/4}$. What is the error?

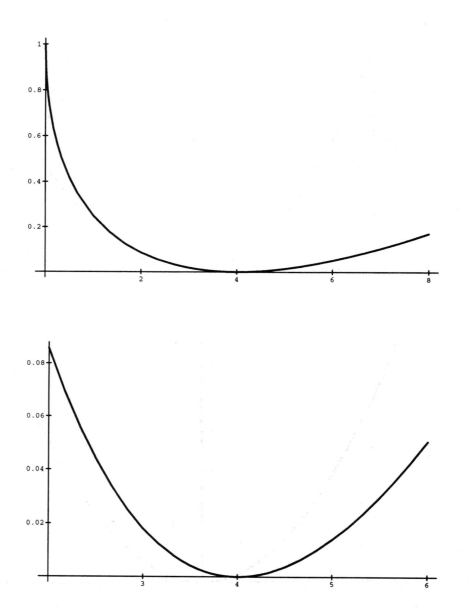

Figure 4.2: The absolute error of tangent line approximations to $y = \sqrt{x}$ on large intervals about $x = 4$.

4.1. LINEAR APPROXIMATION

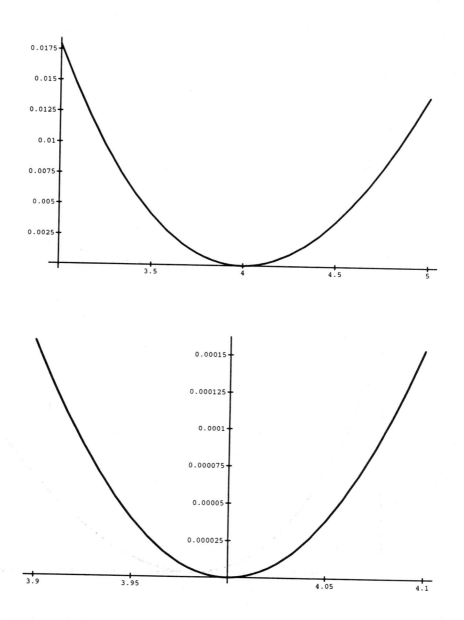

Figure 4.3: The absolute error of tangent line approximations to $y = \sqrt{x}$ on small intervals about $x = 4$.

4. For $k = 1/2, 1/3$ and $1/4$ find the largest interval $[-a, a]$ such that the linear approximation of Exercise 3 has absolute error at most 0.01 for all x in $[-a, a]$.

5. What is the linear approximation formula for $\sin x$ near $x = 0$? What is the largest interval $[-a, a]$ such that this linear approximation has absolute error at most 0.01?

6. Repeat Exercise 5 for $f(x) = \dfrac{5 - 6x + x^2}{(1 + x^2)^{1/3}}$.

7. Consider the function $f(x) = x^{1/3}$ on the interval $[0, 1]$.

    (a) Use errorplot to find the maximum error of the linear approximation to f near $x = 0.1$ on the interval $[0, 0.2]$.

    (b) Repeat part a) for linear approximations to f near $x = 0.3, 0.5, 0.7$ and 0.9 on the intervals $[0.2, 0.4], [0.4, 0.6], [0.6, 0.8]$ and $[0.8, 1]$, respectively. What is the maximum absolute error of these linear approximations over the interval $[0, 1]$? (One nice way to see this is to use errorplot on each subinterval, and then use Show to combine all five graphs into one.)

    (c) Suppose now the interval $[0, 1]$ is divided into n subintervals of equal length, and that on each such subinterval the linear approximation to f is obtained near the midpoint of the subinterval. (This was done for the case $n = 5$ in parts a) and b).) What is the least value of n which will ensure that the maximumu absolute error over the n subintervals is less than 0.01?

    (d) Instead of dividing $[0, 1]$ into subintervals of *equal* length, suppose you can pick any n subintervals of $[0, 1]$, but that you still use the linear approximation to f near the midpoint of each of these intervals. What is the least value of n which will ensure that the maximum absolute error over the n subintervals is less than 0.01?

4.2 The Bisection Method

If f is continuous on the interval $[a, b]$ and $f(a)f(b) < 0$ — that is, f changes sign on the interval — then there is a number c in the interval $[a, b]$ such that $f(c) = 0$. (See p 62, Edwards and Penney[3].) This consequence of the Intermediate Value Property suggests a method for approximating solutions to the equation

$$f(x) = 0,$$

known as the *Method of Bisection*.

4.2. THE BISECTION METHOD

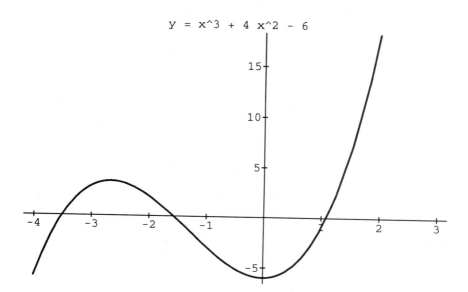

Figure 4.4: There are three real solutions to the equation $x^3 + 4x^2 - 6 = 0$

Example 1. Consider the equation $x^3 + 4x^2 - 6 = 0$. This cubic cannot easily be factored so finding exact solutions is incovenient. The most practical approach is to *approximate* them. Let

$$f(x) = x^3 + 4x^2 - 6.$$

If you plot its graph, Figure 4.4, you can see that there are three solutions to the equation $f(x) = 0$, one in each of the intervals $[-4, -3], [-2, -1]$ and $[1, 2]$.

```
In[1]:= f[x_] := x^3 + 4 x^2 - 6
In[2]:= Plot[f[x], {x, -4, 3}];
```

Let us concentrate on the solution to $f(x) = 0$ in the interval $[-2, -1]$. The graph of $y = f(x)$ obviously crosses the x-axis somewhere between -2 and -1. In which half of $[-2, -1]$ does this x-intercept occur? To see this, we can plot f on the interval $[-2, -1]$.

```
In[3]:= Plot[f[x], {x, -2, -1}];
```

From the graph, see Figure 4.5, it is obvious that the root is in the *left* half of $[-2, -1]$. The method of bisection simply repeats this process over and over: find which half of a given interval the root is in. Every time we bisect the interval, we have isolated the root in an interval only half as long as the previous one. This way we get closer and closer to the actual root:

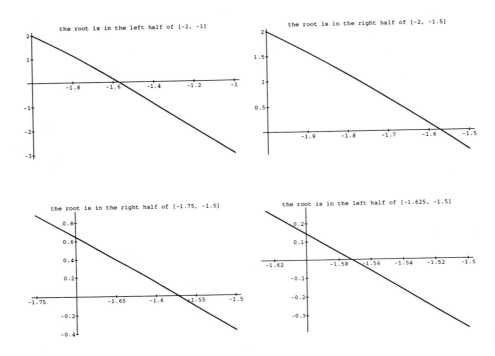

Figure 4.5: Four applications of the Bisection Method.

```
In[4]:= Plot[f[x], {x, -2, -1.5}];
In[5]:= Plot[f[x], {x, -1.75, -1.5}];
In[6]:= Plot[f[x], {x, -1.625, -1.5}];
```

Thus after four bisections of $[-2, -1]$ we know the root is somewhere between -1.625 and -1.5625.

Example 2. So far we have only been using graphs to focus in on the solution to $x^3 + 4x^2 - 6 = 0$ in $[-2, -1]$. To obtain a numerical approximation to the root, simply take the midpoint of the latest interval in which the root must lie: after the four bisections of Example 1 we could say the root is approximately -1.59375. However, since *Mathematica* is doing all its calcuations to six–digit precision, we could simply keep bisecting the interval until the left and right endpoints are the same (to six digits.) Then we could take this common value as our approximation. There is a defined function in `Chap4.m` to do the calculations of the bisection method for you, namely `bisect`:

```
In[7]:= << Chap4.m    (* Out[7] will be a long usage
```

4.2. THE BISECTION METHOD

```
In[8]:= ?bisect
bisect[f][{a, b}] determines which half of the interval {a, b}
contains a solution to the equation f[x] == 0. It is assumed
that f is continuous on {a, b}. If f[a] and f[b] have the same
sign, then the output is the original interval.
```

Note that `bisect[f]` acts on ordered pairs and returns ordered pairs.

```
In[8]:= bisect[f][{-2, -1}]  (* f as before *)
Out[8]= {-2., -1.5}
In[9]:= bisect[f][%]
Out[9]= {-1.75, -1.5}
In[10]:= bisect[f][%]
Out[10]= {-1.625, -1.5}
In[11]:= bisect[f][%]
Out[11]= {-1.625, -1.5625}
```

This process could be continued and would result in better approximations. However, to simplify calculations, we can make use of *Mathematica's* built-in function `Nest`:

```
In[12]:= ?Nest
Nest[f, expr, n] gives an expression with f applied n times to expr.
```

Thus `Nest` can be used to perform repeated compositions of a function with itself.

Example 3. We can repeat all the calculations of Example 2 in one line using `Nest`, and then we can proceed to apply the bisection method repeatedly until the interval stabilizes:

```
In[12]:= Nest[bisect[f], {-2, -1}, 4]
Out[12]= {-1.625, -1.5625}
In[13]:= Nest[bisect[f], {-2, -1}, 10]
Out[13]= {-1.57227, -1.57129}
In[14]:= Nest[bisect[f], {-2, -1}, 15]
Out[14]= {-1.57202, -1.57199}
In[15]:= Nest[bisect[f], {-2, -1}, 16]
Out[15]= {-1.57201, -1.57199}
In[16]:= Nest[bisect[f], {-2, -1}, 17]
Out[16]= {-1.572, -1.57199}
In[17]:= Nest[bisect[f], {-2, -1}, 18]
Out[17]= {-1.57199, -1.57199}
```

Thus, to six digits, one solution to $x^3 + 4x^2 - 6 = 0$ is -1.57199.

Example 4. The fact that 18 bisections are required to approximate the solution in $[-2, -1]$ to 6 digits is not a coincidence. Each time we bisect an interval we obtain a new interval only half as long as the previous one. After 18 bisections we obtain an interval of length

$$(\frac{1}{2})^{18} \simeq 0.0000038,$$

since we started with an interval of length one. Therefore the 18*th* bisection will result in an interval with left and right end points differing by only 0.0000038. Since *Mathematica* works, by default, with six–digit precision, after round off, the end points of the 18*th* interval can differ by at most 0.00001. So to obtain the other two roots of the equation $f(x) = 0$ to six digits, we simply need to apply **bisect** 18, possibly 19 or 20, times to each of the intervals $[-4, -3]$ and $[1, 2]$.

```
In[19]:= Nest[bisect[f], {-4, -3}, 18]
Out[19]= {-3.51414, -3.51413}
```

(One more iteration is required.)

```
In[20]:= Nest[bisect[f], {-4, -3}, 19]
Out[20]= {-3.51414, -3.51414}
In[21]:= Nest[bisect[f], {1, 2}, 18]
Out[21]= {1.08613, 1.08613}
```

Thus, to six digits, the three roots of

$$x^3 + 4x^2 - 6 = 0$$

are -3.51414, -1.57199 and 1.08613.

Example 5. In light of the above calculations, which are very systematic, it should come as no surprise that *Mathematica* has a built–in function, **NRoots**, to approximate roots of a polynomial equation.

```
In[22]:= ?NRoots
NRoots[lhs==rhs, var] gives a list of numerical approximations to
the roots of a polynomial equation.\index{roots}
```

If we apply **NRoots** to the polynomial $f(x)$ we can get approximations to all three roots at once.

```
In[23]:= NRoots[x^3 + 4 x^2 - 6 == 0, x]
Out[23]= x == -3.51414 || x == -1.57199 || x == 1.08613
```

4.2. THE BISECTION METHOD

> **MATHEMATICA FUNCTIONS INTRODUCED IN THIS SECTION**
> Nest   NRoots

Exercises:

1. Determine intervals in which the equation $x^3 - 4x + 1 = 0$ has its solutions, and then use the method of bisection to approximate each root to six decimal places. Compare your answers with the results obtained by using `NRoots`.

2. Repeat Exercise 1 for the equation $x^5 + x = 1$. Note that `NRoots` finds all five solutions whereas the Method of Bisection finds only the *real* roots.

3. Consider the equation $x^3 + 45x^2 - 574x - 4007 = 0$. Plot the graph of

$$f(x) = x^3 + 45x^2 - 574x - 4007$$

   to find intervals in which the real solutions to $f(x) = 0$ lie. Use the method of bisection to approximate each real solution to six digits. (NB: If your initial interval has length greater than one, you will have to use more than 20 iterations of the bisection method.) Compare with the results obtained by using `NRoots`. Also, compare with the results obtained by using `Solve[f(x) == 0, x] // N`.—What do you notice?

4. Consider the equation

$$x - 2^{-x} = 0.$$

   Approximate all real solutions to this equation using the bisection method. Try using `NRoots`; why doesn't it work? Try using `Solve`. What happens?

5. Repeat Exercise 4 for the equation

$$e^x + 2^{-x} + 2\cos x - 6 = 0.$$

6. Consider the equation $\sin(1/x) = 0$. Let $f(x) = \sin(1/x)$; plot the graph of f on the interval $[-1, 1]$. How many real solutions are there to the equation $f(x) = 0$? Use the method of bisection to approximate the first five solutions between 0 and 1, counting from the *right* side of the interval.

7. Actually Exercise 6 is a bit of a cheat since it is quite easy to find the *exact* solutions to $\sin(1/x) = 0$. What are the exact values of the five solutions you approximated in Exercise 6? (But try solving for the exact solutions with `Solve`! Interrupt *Mathematica* if it hasn't produced a solution within 5 minutes.)

8. We have been doing all our calculations with six–digit precision, but of course *Mathematica* can work with more precision if you intstruct it to. Thus in theory the method of bisection could be used to find approximations to any desired degree of precision — but the number of iterations required will increase. Condsider the general case of a continuous function f defined on an interval $[a_0, b_0]$ such that
$$f(a_0)f(b_0) < 0.$$
Suppose the method of bisection is used to find a solution to the equation $f(x) = 0$ in the interval $[a_0, b_0]$, and that after n bisections we know there is a solution in the interval $[a_n, b_n]$. Finally, suppose that at this stage we take the midpoint of $[a_n, b_n]$, call it p_n, as our approximation to the solution. Thus
$$p_n = \frac{a_n + b_n}{2}.$$

$$a_0 \qquad\qquad a_n\ p_n\ b_n \qquad\qquad\qquad\qquad b_0$$

Suppose the actual root we are approximating is r; so r is in the interval $[a_n, p_n]$ or in the interval $[p_n, b_n]$. Prove that
$$|p_n - r| \leq \frac{1}{2^{n+1}}(b_0 - a_0).$$
What is the least value of n that will ensure that
$$|p_n - r| \leq 0.0000000001 = 10^{-10}?$$
What is this value of n if $b_0 - a_0 = 1$?

4.3 Newton's Method

The method of bisection is a very powerful method for approximating solutions to equations of the form
$$f(x) = 0.$$
It is simple to use, and the only restriction on f is that it be continuous on $[a, b]$. However, as we saw in the last section, the number of iterations required can be quite large. This causes no difficulty if you have access to a computer; nevertheless, it is only natural to wonder if there are other approximation methods which require fewer iterations than the bisection method. If we add the condition that f also be differentiable, then it is possible to approximate solutions to the equation $f(x) = 0$

4.3. NEWTON'S METHOD

by another method known as *Newton's Method*. The key to Newton's method is the observation that if x_0 is close to the desired root, then the x-intercept of the tangent to f at $x = x_0$, call it x_1, is *closer* to the desired root. (See Figure 4.6.) Repeating this process we can obtain a sequence of approximations,

$$x_0, x_1, x_2, \ldots, x_n, x_{n+1}, \ldots$$

to the desired root. The iteration formula for x_{n+1}, for $n \geq 0$, is

(4.1) $$x_{n+1} = x_n - \frac{f(x_n)}{f'(x_n)},$$

Two things should be pointed out about equation 4.1:

- The initial choice of x_0 should be made with care — the closer to the desired root the better.

- The sequence of values $x_1, x_2, \ldots, x_n, \ldots$ *may* not converge.

However, when Newton's method does work, it works so well that the convergence is usually much quicker than that of the bisection method.

Example 1. In the package for Chapter 4 there is a function defined to exhibit Newton's method. It is called `newtonsmethod`.

```
In[1]:= << Chap4.m
In[2]:= ?newtonsmethod
newtonsmethod[f, a, {b, c}][n] shows Newton's method being used to
generate the (n + 1)st approximation to the solution of the equation
f[x] == 0 from the nth approximation, starting with initial guess
x0 = a. All previous approximations are shown along the x-axis,
and the value of the current approximation is indicated in the
PlotLabel. The interval {b, c} is the initial x-interval used for the
plot; however the x-interval will adjust automatically, if necessary,
to include all approximations x0, x1, ... , xn,x(n+1).
```

As an example, let us consider once again the equation

$$x^3 + 4x^2 - 6 = 0.$$

As an initial guess let us use $x_0 = 0.4$, which is not a particularly good first choice, but therefore the convergence to the root in the interval $[1, 2]$ will take more steps so we can see better what is happening. The first four following outputs are on display in Figure 4.6 and Figure 4.7.

116   CHAPTER 4. METHODS OF APPROXIMATION

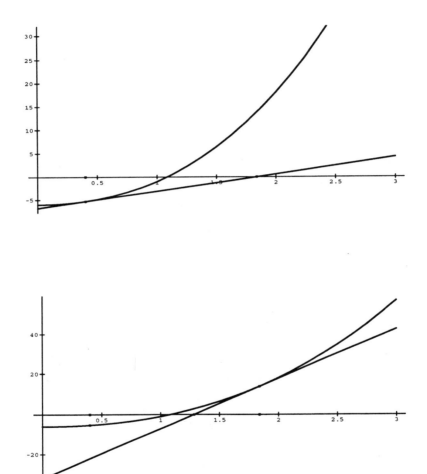

Figure 4.6: Beginning Newton's Method. Top: initial guess 0.4; Bottom: after 2 iterations, approximation is 1.28603.

4.3. NEWTON'S METHOD

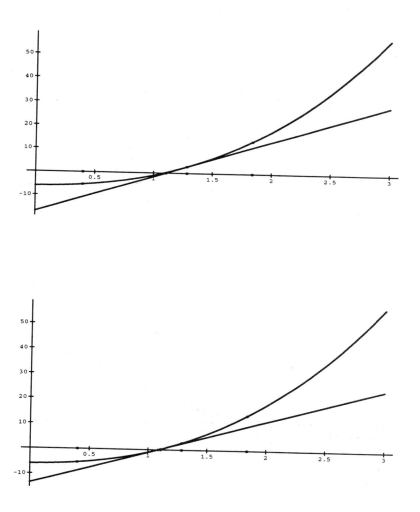

Figure 4.7: Further iterations of Newton's Method. Top: after 3 iterations approximation is 1.1062; Bottom: after 4 iterations approximation is 1.08636.

CHAPTER 4. METHODS OF APPROXIMATION

```
In[2]:= f[x_] := x^3 + 4 x^2 - 6
In[3]:= newtonsmethod[f, 0.4, {0, 3}][0];
In[4]:= newtonsmethod[f, 0.4, {0, 3}][1];
In[5]:= newtonsmethod[f, 0.4, {0, 3}][2];
In[6]:= newtonsmethod[f, 0.4, {0, 3}][3];
In[7]:= newtonsmethod[f, 0.4, {0, 3}][4];
```

Note that each graph produced with `newtonsmethod` keeps track of all the previous approximations, and gives you the value of the current approximation. Of course, if the initial guess were closer to the root then the convergence would be quicker. Consider what happens if we take $x_0 = 1.5$ and look at the 4th iteration:

```
In[8]:= newtonsmethod[f, 1.5, {0, 3}][3];
```

In this case you can see that it requires only 4 iterations of Newton's method to obtain the root in $[1, 2]$ to six digits — namely 1.08613. This is much less than the 18 iterations required by the bisection method.

Example 2. The iteration formula of equation 4.1 is also defined in the package `Chap4.m`. It is called `iteration`.

```
In[9]:= ?iteration
iteration
                                       f[x]
iteration/: iteration[f_][x_] := x - -----
                                      f'[x]
```

In terms of `iteration`, Equation 4.1 can be written as

$$x_{n+1} = \text{iteration}[f](x_n).$$

Thus you can use `iteration` along with `Nest` if you wish to simply compute approximations without viewing the process graphically.

```
In[9]:= Nest[iteration[f], 0.4, 6]    (* f as before *)
Out[9]= 1.08613
In[10]:= Nest[iteration[f], 1.5, 2]
Out[10]= 1.08916
In[11]:= Nest[iteration[f], 1.5, 3]
Out[11]= 1.08614
```

Here the approximations are all getting closer to the root in $[1, 2]$. But if we pick a negative initial choice, Newton's method will converge to one of the other two roots.

4.3. NEWTON'S METHOD

```
In[12]:= Nest[iteration[f], -1.5, 3]
Out[12]= -1.57199
In[13]:= Nest[iteration[f], -1.5, 4]
Out[13]= -1.57199
In[14]:= Nest[iteration[f], -20, 4]
Out[14]= -(201266870759680151356684361080252571078354228561942974212
3415629486276586140349547069448800839268573743326543919369054336550
5664572234129370583170251596600157132903114982087181158204886645878804955)
```

Note the blunder in Input[14]: the initial guess was entered as $x_0 = -20$ *without* a decimal point. Thus *Mathematica* gives the exact result of applying 4 iterations of Newton's Method to $x_0 = -20$. Even though the fraction has many, many digits, the interesting thing to observe is that the number in Out[14] is a *rational* number, and thus could have been calculated by hand — even though you would probably never want to do it! However, to get a decimal approximation, we can simply enter our initial guess as -20. *with* a decimal point, and then *Mathematica* will respond with same:

```
In[15]:= Nest[iteration[f], -20., 4]
Out[15]= -5.37668
In[16]:= Nest[iteration[f], -20., 8]
Out[16]= -3.51504
```

Note that even though $x_0 = -20$ is a fairly bad first approximation to the root with value -3.51414, after eight iterations Newton's method is already very close to the actual root.

Example 3. Newton's method will not work if x_0, or any later x_n for that matter, is a critical point of f. You can see why if you consider Equation 4.1: you would be dividing by zero. Geometrically, if $f'(x_0) = 0$, it means the tangent to f at $x = x_0$ is *parallel* to the x-axis and will never intersect it. For f as above, one critical point is 0. If you try $x_0 = 0$ as an initial guess you will not get a numerical answer: *Mathematica* will simply give you a warning:

```
In[17]:= Nest[iteration[f], 0.0, 4]
                                    1
Power::infy: Infinite expression  --  encountered.
                                    0.
Infinity::indt:
   Indeterminate expression ComplexInfinity +
ComplexInfinity encountered.
Infinity::indt:
   Indeterminate expression ComplexInfinity +
```

ComplexInfinity encountered.
Out[17]= Indeterminate

Example 4. Newton's method may converge to a root of $f(x) = 0$ even if you make a *terrible* choice for x_0; it will just take longer to get close to the root. For instance, if $x_0 = 400$ it will take 20 iterations to obtain the root 1.08613.

In[18]:= Nest[iteration[f], 400., 20]
Out[18]= 1.08613

Example 5. *Mathematica* offers a function, **FindRoot**, which gives numerical approximations to solutions of an equation, depending on an initial input.

In[19]:= ?FindRoot
FindRoot[lhs==rhs, {x, x0}] searches for a numerical solution to the equation lhs==rhs, starting with x=x0.

We can use **FindRoot** to approximate the three solutions to

$$x^3 + 4x^2 - 6 = 0$$

by entering three different suitable initial guesses:

In[19]:= FindRoot[x^3 + 4 x^2 - 6 == 0, {x, 1.5}]
Out[19]= {x -> 1.08613}
In[20]:= FindRoot[x^3 + 4 x^2 - 6 == 0, {x, -1.5}]
Out[20]= {x -> -1.57199}
In[21]:= FindRoot[x^3 + 4 x^2 - 6 == 0, {x, -3.5}]
Out[22]= {x -> -3.51414}

If we try using **FindRoot** to solve the equation $\sin x = x$, which has obvious solution $x = 0$, but we input an initial guess of $x_0 = 12$ something interesting happens:

In[23]:= FindRoot[Sin[x] == x, {x, 12}]
FindRoot::convNewt:
    Newton's method failed to converge to the prescribed accuracy after 15 iterations.
Out[23]= {x -> -2.76651 10^7}

Mathematica's message is very revealing — **FindRoot** uses Newton's method. In this case, it would be applying Newton's method to the function $f(x) = \sin x - x$. The reason Newton's method fails to converge quickly is obvious if you plot the

4.3. NEWTON'S METHOD

graph of f, or view the first few iterations using `newtonsmethod`. (Try it!) The graph of f simply "wiggles" too much and its tangents vary wildly in direction. `FindRoot`'s test for convergence is to substitute x_n into the equation to see if $lhs = rhs$ to six–digit precision. If not, it tests x_{n+1}. By default, `FindRoot` will test the first 15 iterations of Newton's method this way. If no acceptable solution has been found by then, it will present you the above message, and the result of doing the 15th iteration — in this case, `Out[23]`. But the value in `Out[23]` is somewhat of a mystery! In theory we could check it: let $f(x) = \sin x - x$, let $x_0 = 12$, and compute x_{15}. However, doing this gives a totally different result than `Out[23]`:[1]

```
In[24]:= f[x_] := Sin[x] - x
In[25]:= Nest[iteration[f], 12., 15]
Out[26]= 314.438
```

You can try 14, 16, 17, or 18 iterations, but neither will result in an answer remotely resembling `Out[23]`! However, 50 iterations will get you close to the root:

```
In[27]:= Nest[iteration[f], 12., 50]
Out[28]= 0.0000200524
```

Example 6. By considering each output of `newtonsmethod[f, a, {b, c}][n]` for different values of n as a separate "frame", you can in DOS or on a NeXT animate a list of such frames (eg by using `Animate` from the animation package `Animation.m`) to form a "movie" of Newton's method in action. An example of such an animation file is available, *newtons.ani*, for you to run through. You can create your own animations frame by frame — which is often the best way, since you may wish to standardize the PlotRange or other options for each frame —, or you can simply do it all at once with a command like

`Animate[Release[Table[newtonsmethod[f, a, {b, c}][i], {i, 0, n}]]]`

which will create a movie of frames 0 to n.

MATHEMATICA FUNCTIONS INTRODUCED IN THIS SECTION
FindRoot

Exercises:

1. You can prove the iteration formula of Equation 4.1 using *Mathematica*, but you have to use different names for your variables, since *Mathematica* input

---

[1] It probably has to do with round-off error in the intermediate calculations — `FindRoot` is probably working with a different degree of precision than *Mathematica*'s default setting.

does not accept subscripts. Let $old = x_n$, let $new = x_{n+1}$. Use Solve and tangent to prove
$$new = old - \frac{f(old)}{f'(old)}.$$

2. Consider the equation $x^3 - 2x - 5 = 0$. How many real solutions are there to this equation? Approximate its root(s) to six digits using Newton's method. Use newtonsmethod to view the results: start with $x_0 = 3$ and plot on the interval $[1.5, 3.2]$. How many iterations are required until the approximation (to six digits) no longer changes?

3. Use FindRoot to confirm the result of Exercise 2.

4. Consider the equation $3x^2 - e^x = 0$. How many real solutions are there to this equation? Use Newton's method to approximate all the real roots to six digits.

5. Confirm the results of Exercise 4 with FindRoot.

6. Consider the function
$$f(x) = \frac{4x - 7}{x - 2}.$$
What is the exact solution to $f(x) = 0$? Use Newton's method to approximate the solution of $f(x) = 0$; start with the following initial values and use newtonsmethod to explain the results:

   (a) $x_0 = 1.6125$
   (b) $x_0 = 1.875$
   (c) $x_0 = 1.5$
   (d) $x_0 = 3.0$

7. In this exercise we shall try Newton's method with various different initial choices to approximate solutions to the equation $x + \tan x = 0$. (A solution to this equation is simply an intersection point of $y = -x$ and $y = \tan x$, so you may wish to plot these two graphs, say on $[-10, 10]$, to get a rough idea of where the solutions are.) Let $f(x) = x + \tan x$ and try initial choices $x_0 = 1, 11$, and 24. Obtain your approximations to six digits. Compare your computations with the results given by FindRoot. Why is the convergence so slow if $x_0 = 24$? Does Newton's method even converge to a solution of $f(x) = 0$ if $x_0 = 24$? Why, or why not?

8. As Exercise 7 and Example 3 show, Newton's method is not always the best way to approximate solutions to an equation of the form $f(x) = 0$. Thus, in some instances, the method of bisection should be used. Use it to approximate (to six digits) all the solutions of $x + \tan x = 0$ in the interval $[-10, 10]$. (There are seven of them, one of which is the obvious one, $x = 0$.)

4.4. FIXED POINT ITERATION

9. Let $f(x) = x^2 - a$. Find an iteration formula in terms of f that can be used to approximate $\sqrt{a}$.

10. Use Newton's method to approximate $\sqrt{2}$ correct to six digits.

11. Let $f(x) = x^3 - a$. Find an iteration formula in terms of f that can be used to approximate $\sqrt[3]{a}$.

12. Use Newton's method to approximate $\sqrt[3]{25}$ to six digits.

13. Using any method you wish, find to six digits all the real roots of the equation
$$\frac{1}{2} + \sin x = \frac{x}{3}.$$

14. What are the exact solutions to $(x+1)(x-1)^3 = 0$? Use Newton's method with initial guesses -1.5 and 1.5 to approximate each root to six digits. How many iterations are required if $x_0 = -1.5$? How many iterations are required if $x_0 = 1.5$? How can you explain the difference?

4.4 Fixed Point Iteration

Consider the equation $\cos x = x$. Any solution to this equation is an intersection point of $y = \cos x$ and $y = x$. If you plot these two graphs you will see that there is one intersection point between 0 and 1.

```
In[1]:= Plot[{Cos[x], x}, {x, -0.5, 2}];
```

Example 1. We could use any of the previous approximation methods to find this root to six digits. Suppose we use the method of bisection:

```
In[2]:= f[x_] := Cos[x] - x
In[3]:= << Chap4.m
In[4]:= Nest[bisect, {0, 1}, 21]
Out[4]= {0.739085, 0.739085}
```

After 21 iterations we find the root is approximately 0.739085. What is interesting about this example is that there is another way to find this approximation, and it simlpy involves repeated calculations with the cosine function. Suppose we start with $x = 0$. The first few iterations are done in the following calculations.

```
In[5]:= Cos[0.0]
Out[5]= 1.
```

```
In[6] := Cos[%]
Out[6] = 0.540302
In[7] := Cos[%]
Out[7] = 0.857553
In[8] := Nest[Cos, 0.0, 10]
Out[8] = 0.731404
```

If we define $x_0 = 0$ and $x_{n+1} = \cos x_n$, then we can follow the results of our first few calculations in Figure 4.8 and Figure 4.9. We start with the point $(x_0, x_1) = (0, 1)$ on the graph of $y = \cos x$ and from it move to the point $(x_1, x_1) = (1, 1)$ on the graph of $y = x$. Since $x_2 = \cos x_1$, the point (x_1, x_2) is directly below the point (x_1, x_1), on the curve $y = \cos x$. Then from the graph of $y = \cos x$ we move directly over to the line $y = x$, to the point (x_2, x_2). Continuing in this fashion we can see that the "cobweb" joining successive points is "homing in" on the intersection point of $y = \cos x$ and $y = x$. Thus repeated iterations of the cosine function should eventually produce a good approximation to the root of $\cos x = x$. We can calcuate further values by applying Nest to the cosine function:

```
In[9] := Nest[Cos, 0.0, 20]
Out[9] = 0.738938
In[10] := Nest[Cos, 0.0, 30]
Out[10] = 0.739082
In[11] := Nest[Cos, 0.0, 40]
Out[12] = 0.739085
```

Finally, after 40 iterations we obtain the correct approximation, to six digits.

In general, any solution p to the equation

$$g(x) = x$$

is called a *fixed point* of g, and the iteration formula

$$x_{n+1} = g(x_n)$$

is referred to as *Fixed Point Iteration*. In many cases — depending on the function g and the choice of x_0 — the sequence of x_n will actually converge to a fixed point of g.

Example 2. There is a defined function in Chap4.m, called fixedpoint which will allow you to follow graphically the results of trying to find a fixed point of a function by using fixed point iteration. It was used to create the graphs of Figures 4.8 and 4.9, as follows:

```
In[13] := fixedpoint[Cos, 0, {-0.5, 2}][0];
```

4.4. FIXED POINT ITERATION

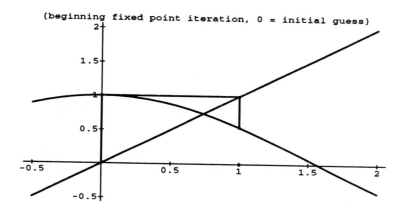

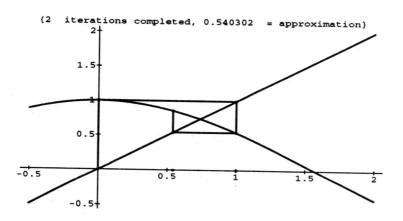

Figure 4.8: Beginning Fixed Point Iteration to solve the equation $\cos x = x$.

CHAPTER 4. METHODS OF APPROXIMATION

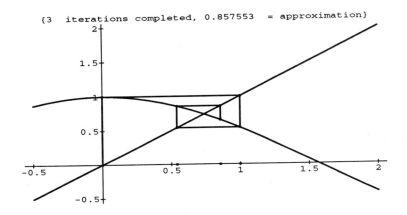

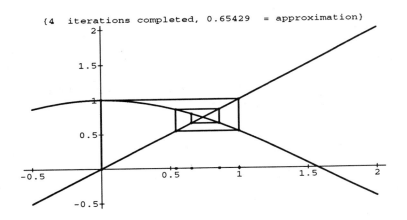

Figure 4.9: Further iterations to find the fixed point of $\cos x$.

4.4. FIXED POINT ITERATION

```
In[14]:= fixedpoint[Cos, 0, {-0.5, 2}][1];
In[15]:= fixedpoint[Cos, 0, {-0.5, 2}][2];
In[16]:= fixedpoint[Cos, 0, {-0.5, 2}][3];
```

Example 3. *Mathematica* also offers a function, called `FixedPoint`, which allows you to calculate fixed points of functions:

```
In[17]:= ?FixedPoint
FixedPoint[f, expr] starts with expr, then generates a sequence of
expressions by repeatedly applying f, until the result no longer
changes. FixedPoint[f, expr, n] stops after at most n steps.
In[17]:= FixedPoint[Cos, 0., 10]  (* compare with In[8] *)
Out[17]= 0.731404
```

If no limit on the number of iterations is specified, `FixedPoint` will continue iterating until a fixed point is obtained:

```
In[18]:= FixedPoint[Cos, 0.0]
Out[18]= 0.739085
```

Warning: if no fixed point exists, and no maximum number of iterations is specified, *Mathematica* may calculate a long time, and may need to be interrupted!

Example 4. It should be pointed out that the iteration formula for Newton's method, Equation 4.1, invloves fixed point iteration on the function $x - \dfrac{f(x)}{f'(x)}$. Thus using `FixedPoint` on `iteration[f]` is equivalent to using Newton's method to solve the equation $f(x) = 0$.

```
In[19]:= FixedPoint[iteration[f], 0.5]  (* f[x] = Cos[x] - x, still *)
Out[19]= 0.739085
```

Example 5. You can make animations of fixed point iteration in action just as you did with Newton's method. There is an example of this on file, *fixedpt.ani*, which you should run through. In Chapter 13 we return to fixed point iteration and describe how you can use it to investigate chaos.

MATHEMATICA FUNCTIONS INTRODUCED IN THIS SECTION
FixedPoint

FUNCTIONS CONTAINED IN THE PACKAGE Chap4.m		
bisect	iteration	normal
tangent	tangentplot	errorplot
newtonsmethod	fixedpoint	normalplot

Exercises:

1. Use `fixedpoint` to view the first few steps of the fixed point iteration method applied to the equation $\sin x = x$. Start with an initial guess of 0.5. What is the (only) real solution to this equation?

2. Repeat Exercise 1 for the equation $\sqrt{x} = x$, starting with $x_0 = 1.4$.

3. Repeat Example 2, but start with a smaller initial plot interval, say $[0.5, 0.9]$, take $x_0 = 0.6$, and take $n = 10$. This will allow you to focus in on the action near the intersection point.

4. Let $f(x) = x^2 - 2$ and let $g(x) = \text{iteration}[f](x)$, the function obtained from Newton's method applied to the equation $f(x) = 0$. Repeat Exercise 1 for g, starting with an initial guess of $x_0 = 2$. Compare the "rate of convergence" of the calculations in this exercise with those from Exercises 1 and 2.

5. If p is a fixed point of g, use the results of Exercises 1, 2, and 4 to make a conjecture, in terms of $g'(p)$, as to when fixed point iteration converges "quickly" or "slowly".

6. Reformulate your conjecture from Exercise 5 for Newton's method: if p is a solution to the equation $f(x) = 0$, in terms of $f'(p)$, when does Newton's method converge "quickly" or "slowly"? Compare with Exercise 14 of Section 4.3.

7. Conceptually, fixed point iteration is a very simple method, but to apply it to a general equation of the form $f(x) = 0$ to find a solution p involves finding a function g such that
$$g(p) = p \Leftrightarrow f(p) = 0.$$
(Of course, you could always use Newton's method to find such a function g, as in Exercise 4 above, but then you need to calculate the derivative of f — and derivatives can get very complicated.) Many times more than one such g can be found, but the choice of g can effect drastically the rate of convergence of fixed point iteration, or whether it even converges or not! Consider the following equation:
$$x^4 = x + 2.$$
It has a solution in $[1, 2]$. Show (algebraically) that each of the following choices of g has a fixed point p such that
$$g(p) = p \Leftrightarrow p^4 = p + 2.$$

  (a) $g(x) = x^4 - 2$

  (b) $g(x) = \dfrac{x + 2}{x^3}$

4.4. FIXED POINT ITERATION

(c) $g(x) = (x+2)^{1/4}$

For which of these functions, g, does fixed point iteration actually converge to the solution of $x^4 = x + 2$ in $[1, 2]$; for which g does it converge the quickest? Start all your calculations with an initial guess of $x_0 = 1.5$.

8. How many real solutions are there to the equation $x^3 = x$? What are they? What happens if you apply fixed point iteration to $g(x) = x^3$ to approximate the solutions of $x^3 = x$? Apparently, what must x_0 be if fixed point iteration is to converge to either non-zero solution of $x^3 = x$?

9. Example 4 showed that Newton's method is equivalent to fixed point iteration; something similar can be said for the bisection method. Let

$$f(x) = x^3 + 4x^2 - 6.$$

Apply `FixedPoint` to `bisect[f]`, with appropriate choices for the initial interval, to obtain to six digits the three real solutions of $f(x) = 0$.

Chapter 5

Properties of Differentiable Functions

> *The excess of the sum of the angles of a triangle formed on a surface by shortest lines over two right angles is equal to the total curvature of the triangle.—* C.F.Gauss, Superficies Curvas 1827

5.1 The Mean Value Theorem

The Mean Value Theorem, MVT for short, is one of the fundamental theorems of Calculus. It states:

Theorem 5.1 (MVT) *Suppose that the function f is continuous on the closed interval $[a, b]$ and differentiable on the open interval (a, b). Then*

$$\frac{f(b) - f(a)}{b - a} = f'(c)$$

for some number c in (a, b). See Figure 5.1.

Many important consequences follow from this theorem. For one thing, if f is differentiable on an interval I, it is the Mean Value Theorem that allows one to conclude the following:

- If $f'(x) = 0$ for all x in I, then f is *constant* on I.

- If $f'(x) > 0$ for all x in I, then f is *increasing* on I.

- If $f'(x) < 0$ for all x in I, then f is *decreasing* on I.

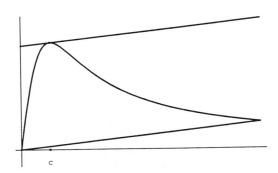

Figure 5.1: The Mean Value Theorem

These results are useful for sketching graphs of functions, and for classifying critical points — maximum, minimum, or neither? — via the First Derivative Test. Similarly significant conclusions can be drawn from the sign of the second derivative:

- If $f''(x) > 0$ for all x in I, then f is *concave up* on I.

- If $f''(x) < 0$ for all x in I, then f is *concave down* on I.

In addition there is a Second Derivative Test for classification of critical points $x = a$ for which $f''(a) = 0$. Even though you are probably already familiar with all of the results above (from your work in high school), the following three examples will use *Mathematica* to illustrate them for you.

Example 1. Consider the polynomial function $f(x) = 1 - 3x + x^3$. Let us find its critical points and inflection points, and then plot

$$f, f', \text{ and } f''$$

all on the same graph.

```
In[1]:= f[x_] := 1 - 3 x + x^3
In[2]:= Solve[f'[x] == 0, x]   (* critical points *)
Out[2]= {{x -> 1}, {x -> -1}}
In[3]:= Solve[f''[x] == 0, x]   (* possible inflection points *)
Out[3]= {{x -> 0}}
In[4]:= Plot[{f[x], f'[x], f''[x]}, {x, -3, 3}]
Out[4]= -Graphics-
```

5.1. THE MEAN VALUE THEOREM

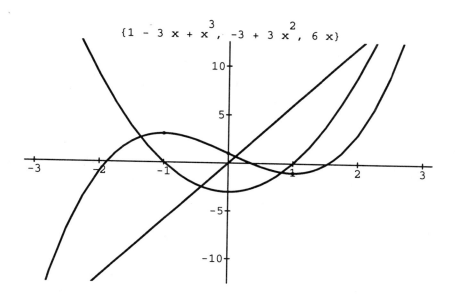

Figure 5.2: The graphs of $y = 1 - 3x + x^3$ and its derivatives.

Looking at this graph is not too illuminating, since we don't know immediately which curve is which. So let us fix things up a bit. First we shall highlight the critical points and possible inflection point:

```
In[5] := ListPlot[{{-1, f[-1]}, {0, f[0]}, {1, f[1]}}];
```

Then we can combine the two previous graphs, adding a label to indicate what the function is, and what its two derivatives are. (See Figure 5.2.)

```
In[6] := Show[%, %%, PlotLabel -> {f[x], f'[x], f''[x]}];
```

Look over Figure 5.2, or your computer screen, if you still have Figure 5.2 on screen. Note that the three highlighted points make it easy to identify the graph of f, and that the formulas for f, f', and f'' across the top of the figure allow you to determine which graph is the first derivative, and which is the second derivative: the parabola is the graph of f'; the line is the graph of f''. Check out the correspodence between the three graphs: f is increasing only when f' is positive; the critical points of f are the zeros of f'; f is concave down only when f'' is negative; etc. Note in particular that the inflection point $(0,1)$ corresponds to the zero of f'', but can also be said to correspond to the *local minimum of f'*.

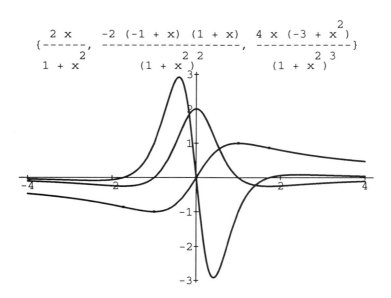

Figure 5.3: The graphs of $y = \dfrac{2x}{1+x^2}$ and its derivatives

Example 2. Let us now repeat the previous example for the rational function

$$g(x) = \frac{2x}{1+x^2}.$$

The sequence of inputs follows; and for the final graphics output of g, g', and g'' see Figure 5.3:

```
In[7]:= g[x_] := 2 x/(1 + x^2)
In[8]:= Solve[g'[x] == 0, x]   (* critical points *)
Out[8]= {{x -> 1}, {x -> -1}}
In[9]:= Solve[g''[x] == 0, x]  (* possible inflection points *)
Out[9]= {{x -> Sqrt[3]}, {x -> -Sqrt[3]}, {x -> 0}}
In[10]:= Plot[{g[x], g'[x], g''[x]}, {x, -4, 4}];
In[11]:= ListPlot[{{-Sqrt[3], g[-Sqrt[3]]}, {-1, g[-1]}, {0, g[0]},
                {1, g[1]}, {Sqrt[3], g[Sqrt[3]]}}];
In[12]:= Simplify[g'[x]]
            -2 (-1 + x) (1 + x)
Out[12]=  --------------------
                   2 2
               (1 + x )
In[13]:= Simplify[g''[x]]
```

5.1. THE MEAN VALUE THEOREM

$$\text{Out[13]} = \frac{4 x (-3 + x^2)}{(1 + x^2)^3}$$

```
In[14]:= Show[Out[10], Out[11], PlotLabel -> {g[x], %%, %}];
```

Look over Figure 5.3. Note that there are three inflection points on the graph of g, namely at $x = -\sqrt{3}$, $x = 0$, and $x = \sqrt{3}$, and that they correspond to the points where g'' crosses the x-axis. In addition you can see that the inflection points of g also correspond to the three extrema of g'. Indeed, it is possible to define an inflection point in terms of the *first derivative* of a function, instead of in terms of its second derivative, which is the usual way. Can you see how to do this? Some exercises below, and the next example, will give you more data to consider.

Example 3. Consider the function h defined by

$$h(x) = x^{5/3} - 5x^{2/3}.$$

This function is defined for all x but none of its derivatives exists at $x = 0$. This will allow us to investigate an example with a critical point, and possibly an inflection point, for which the first derivative and the second derivative respectively, are *not* zero.

```
In[15]:= h[x_] := x^(5/3) - 5 x^(2/3)
In[16]:= Simplify[h'[x]]
          -10 + 5 x
Out[16]= ---------
            1/3
           3 x
In[17]:= Simplify[h''[x]]
          10 + 10 x
Out[17]= ---------
            4/3
           9 x
In[18]:= Plot[{h[x], h'[x], h''[x]}, {x, -6, 6}];
... warnings about nonexistence of h'[0] and h''[0] scroll by ...
In[19]:= ListPlot[{{-1, h[-1]}, {0, h[0]}, {2, h[2]}}];
In[20]:= Show[%, %%, PlotLabel -> {h[x], Out[16], Out[17]}];
```

This graphics output is shown in Figure 5.4, for future reference. In looking at Figure 5.4, focus on the behaviour of all three graphs near $x = 0$. Note that even though $h'(0)$ is not defined, there is still a local maximum on the graph of h at $(0,0)$, as the First Derivative Test will confirm. However, there is no inflection point on the graph of h at $x = 0$ since $h''(x) > 0$ for x near to 0, on either side of 0.

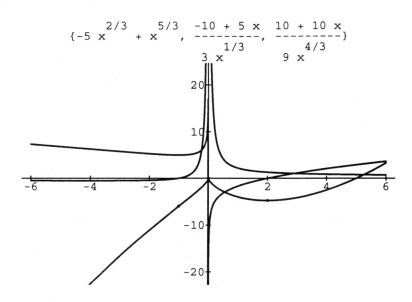

Figure 5.4: The graphs of $y = x^{5/3} - 5x^{2/3}$ and its derivatives

There are many other important consequences of The Mean Value Theorem — L'Hôpital's Rule and The Fundamental Theorem of Calculus, to name only two, can both be proved using MVT. In addition, Rolle's Theorem (which is usually used to prove MVT), can itself be proved using MVT: Rolle's Theorem is just the special case of MVT for which $f(a) = f(b)$. This means that The Mean Value Theorem and Rolle's Theorem are equivalent: one is true if and only if the other is true. This is not surprising at all, if you think about these two theorems in terms of their graphical representations: you can transform the picture for one theorem into that of the other by a simple rotation. There is an animation file, *mvtrt.ani*, which illustrates this point; you should run through it, if your work station supports animation, or use it to construct sequences of images.

Exercises:

1. Consider the quadratic function $f(x) = s + tx + ux^2$ on the interval $[a, b]$. Find the number c that satisfies the conclusion of MVT; show that

$$f'(c) = \frac{f'(a) + f'(b)}{2}.$$

2. In this exercise we shall consider the question of how many real solutions there are to the cubic equation

$$x^3 + px + q = 0,$$

5.1. THE MEAN VALUE THEOREM

cf. Problems 36 to 39 on p 187 of Edwards and Penney [3]. Define the cubic function $f(x) = x^3 + px + q$.

(a) Calculate
$$\lim_{x \to \infty} f(x) \text{ and } \lim_{x \to -\infty} f(x).$$
Conclude there must be at least one real solution to the equation $f(x) = 0$.

(b) Show that if $p \geq 0$, f has no relative extrema. Conclude that the equation $f(x) = 0$ has exactly one real solution.

(c) Now suppose $p < 0$. Find the two critical points, c_1 and c_2, of f. Let c_1 be the negative critical point; c_2, the positive critical point.

(d) Use the second derivative test to show that f has a local maximum at $x = c_1$ and a local minimum at $x = c_1$.

(e) Show that
$$f(c_1)f(c_2) = \frac{4p^3}{27} + q^2.$$
Conclude $f(c_1)f(c_2)$ has the same sign as the expression $4p^3 + 27q^2$.

(f) Let $D = 4p^3 + 27q^2$. Explain why the equation $f(x) = 0$ will have:
   i. three real solutions if $D < 0$
   ii. two real solutions if $D = 0$
   iii. one real solution if $D > 0$

3. Call in the package Chap5.m. In it is a formula cubicplot[p, q] which will plot the cubic function $f(x) = x^3 + px + q$ for different values of p and q.

(a) Try it for the following pairs of (p, q):
$$(2, 5), (-4, 4), (-3, 1), (-3, 2), (-334, 619).$$

(b) Investigate the graph of $f(x) = x^3 - 3x + q$ for q going from 1 to 3. In particular look at what happens for values of q very close to 2.

(c) Investigate the graph of $f(x) = x^3 + px$ for p going from -1 to 1. In particular look at what happens for values of p very close to 0. Note how the cubic "unfolds" as p becomes negative. Combine graphs for different values of p into one graph with Show.

4. Define the general cubic function
$$g(x) = x^3 + ax^2 + bx + c.$$
Simplify $g(z - a/3)$.

(a) Find the values of p and q, in terms of a, b and c, such that
$$g(z - a/3) = f(z).$$

(b) How many real solutions are there to the cubic equation
$$x^3 - 6x^2 + 5x + 10 = 0?$$

5. Repeat Example 1 for the polynomial function $f(x) = 4x^3 - 3x^4 + 1$.

6. Repeat Example 2 for the rational function $g(x) = \dfrac{x^2 - 1}{(1 + x^2)^2}$.

7. Repeat Example 3 for the function h defined by
$$h(x) = x^{4/3} - 4x^{1/3}.$$

8. Based on the data of Examples 1 to 3, and the three previous exercises (or by pure thought alone), write down a test to decide whether or not the point $(a, f(a))$ is an inflection point of f *in terms of f' and its values or features*.

9. Suppose f is a function such that $f(0) = 10$ and
$$f'(x) = \dfrac{x^2 + 15x + 50}{(1 + x^2)^{1/3}}.$$

Sketch a possible graph of f. (No holds barred: use *Mathematica* and any of its functions; use anything and everything you know.) Indicate all critical and inflection points. You may have to use paper and pencil!

5.2 Curvature

Another way to investigate the graph of a function is in terms of its *curvature*, which is defined as the rate of change of the angle that the tangent makes with the x-axis with respect to the distance along the curve. (See Figure 5.5.) To define this mathematically let us suppose f is a function and let P and Q be the points
$$(x, f(x)) \text{ and } (x + h, f(x + h)),$$
respectively, on the graph of f. Let α_P be the angle the tangent line to f at P makes with the x-axis; let α_Q be the angle the tangent line to f at Q makes with the x-axis; let $\overline{PQ}$ be the distance along the graph of f from P to Q. Let κ, the Greek letter kappa, be the curvature of f at x. Then[1]
$$\kappa = \lim_{h \to 0} \dfrac{\alpha_Q - \alpha_P}{\overline{PQ}}.$$

5.2. CURVATURE

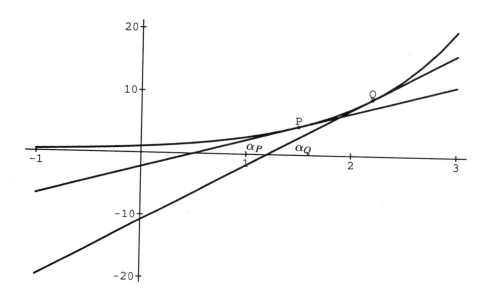

Figure 5.5: Curvature is defined in terms of the angles that tangents make with respect to the x-axis

To evalutate this limit we can use the fact that

$$\tan \alpha_P = f'(x) \text{ and } \tan \alpha_Q = f'(x+h)$$

and that

$$\overline{PQ} \simeq \sqrt{h^2 + (f(x+h) - f(x))^2},$$

since a short segment of a curve is approximately a straight line. We can then use *Mathematica* to evaluate this limit:

```
In[1]:= Limit[(ArcTan[f'[x + h]] - ArcTan[f'[x]])/
            Sqrt[h^2 + (f[x + h] - f[x])^2], h -> 0]
              f''[x]
Out[1]= ----------------
                  2 3/2
            (1 + f'[x] )
```

---

[1] Properly speaking this limit does not exist; can you see why? More commonly, κ is defined as the absolute value of this expression.

Thus[2]
$$\kappa = \frac{f''(x)}{(1+(f'(x))^2)^{3/2}}.$$

There are seven functions defined in `Chap5.m` involving curvature. For instance, there is a function `curvature`:

```
In[2]:= << Chap5.m   (* Out[2] will be a long
                              usage message *)
In[3]:= ?curvature
curvature
                                      (f')'[x]
curvature/: curvature[f_][x_] := ---------------
                                          2 3/2
                                  (1 + f'[x] )
```

Note the form of this defined function: it is defined in terms of the function *f and* a point *x*, and is listable in terms of *x* only. Thus:

```
In[3]:= curvature[Sin][x]
                Sin[x]
Out[3]= -(----------------)
                2 3/2
         (1 + Cos[x] )
In[4]:= curvature[Sin][Pi/2]
Out[4]= -1
```

Example 1. Define a function f such that $f(x) = x^2$. What is its curvature? Plot f and its curvature function. (See Figure 5.6.)

```
In[5]:= f[x_] := x^2
In[6]:= curvature[f][x]
               2
Out[6]= -------------
               2 3/2
        (1 + 4 x )
In[7]:= Plot[{f[x], curvature[f][x]}, {x, -3, 3}];
```

---

[2]More commonly curvature is taken to be positive and thus defined as the absolute value of this expression. But we shall allow positive and negative curvatures: the curvature is positive if the graph of f is concave up; it is negative if the graph of f is concave down. This is then consistent with the case in higher dimensions. For example, a sphere has positive curvature but a saddle has negative curvature; data is still inconclusive on curvatute of spacetime.

5.2. CURVATURE

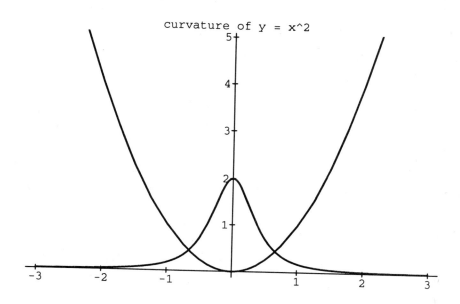

Figure 5.6: The curvature of $y = x^2$

Note that the maximum curvature occurs at $x = 0$ and quickly drops off to almost zero to each side of $x = 0$. To some extent this is obvious if you look at the graph of $y = x^2$: it seems to bend most at the origin.

Chap5.m contains a function to plot the curvature of f along with the graph of f on the interval $[a, b]$. It is called viewcurvature; it is one of many view* functions in Chap5.m.

In[8]:= ?viewcurvature
viewcurvature[f, {a, b}] plots f and its curvature function on the interval {a, b}.

Example 2. Let us investigate the curvature of some well-known curves: a line and a circle. Let $g(x) = mx + b$. What is the curvature of a line? (What do you suppose the answer is?)

In[8]:= g[x_] := m x + b
In[9]:= curvature[g][x]
Out[9]= 0

So a line has zero curvature. What then of a circle? Let

$$h(x) = \sqrt{16 - x^2},$$

the top half of a circle of radius 4.

```
In[10]:= h[x_] := Sqrt[16 - x^2]
In[11]:= curvature[h][x]
```

$$\text{Out[11]}= \frac{-\left(\dfrac{x^2}{(16-x^2)^{3/2}}\right) - \dfrac{1}{\text{Sqrt}[16-x^2]}}{\left(1 + \dfrac{x^2}{16-x^2}\right)^{3/2}}$$

```
In[12]:= Simplify[%]
```

$$\text{Out[12]}= -\left(\dfrac{1}{4}\right)$$

Note that the curvature is constant. The fact that it is negative is because the top of a circle is concave down; the fact that the magnitude is $\frac{1}{4}$ is interesting — the magnitude of the curvature is the reciprocal of the radius. Thus we define the *radius of curvature* to be $|\frac{1}{\kappa}|$. This function is put into `Chap5.m` as well:

```
In[13]:= ?radiusofcurvature
radiusofcurvature

radiusofcurvature/: radiusofcurvature[f_][x_] := Abs[-------------]
                                                     curvature[f][x]
```

For the top half of the circle, defined by h, we find that the radius of curvature is also constant:

```
In[13]:= Simplify[radiusofcurvature[h][x]]
Out[13]= 4
```

In general the curvature along a curve is *not* constant, nor is the radius of curvature. Given a function f and a point x in the domain of f there are *two* points on the normal to f at x which are at a distance equal to the radius of curvature from $(x, f(x))$ — one is below the curve, the other above. Of these two points, one is called the *centre of curvature*, namely the one *above* the curve if f is *concave up*

5.2. CURVATURE

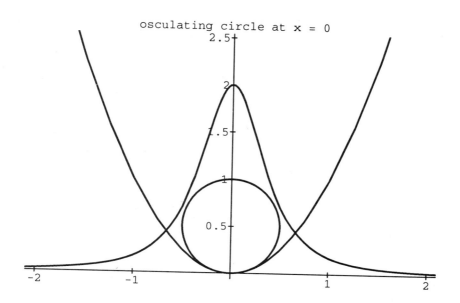

Figure 5.7: The osculating circle of $y = x^2$ at $x = 0$.

at $(x, f(x))$ or the one *below* the curve if f is *concave down* at $(x, f(x))$. Chap5.m includes a function, centreofcurvature, to find the centre of curvature for you:

```
In[14]:= curvature[f][0]    (* recall: f(x) = x^2, still *)
Out[14]= 2
In[15]:= centreofcurvature[f][0]
                1
Out[15]= {0, -}
                2
```

Example 3. The circle with radius equal to the radius of curvature and centre equal to the centre of curvature is called the *osculating circle*. (See Figure 5.7) It is tangent to the curve $y = x^2$ at the point $(0,0)$.

```
In[16]:= viewosculatingcircle[f][0];
In[17]:= Show[%, Out[7]];
```

Compare Figure 5.7 with Figure 5.8, which shows the osculating circle of $y = x^2$ at $x = 1$. Note that in Figure 5.8 the circle is much bigger, since the curvature of f at $x = 1$ is much less. There is an animation file, *osccircl.ani*, showing osculating circles of f moving from left to right along the parabola and through the origin. If

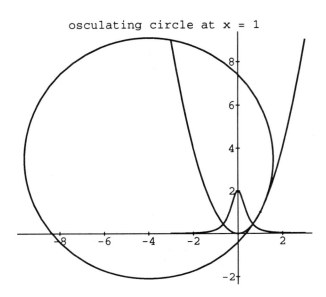

Figure 5.8: The osculating circle of $y = x^2$ at $x = 1$.

your work station supports animation, you should view it. Of course, you can create your own animation files, treating each output of

viewosculatingcircle[f][i]

for different values of the iterator i as a single frame of your animation.

Example 4. As x changes, the centre of curvature changes as well. The locus of centres of curvature forms what is known as the *evolute* of f. There is a defined function in **Chap5.m** to plot the evolute, namely **viewevolute**:

```
In[18]:= ?viewevolute
viewevolute[f, {a, b}] plots the centres of curvatue of f for all x
in the interval {a, b}. This curve is called the evolute of f.
In[18]:= viewevolute[f, {-3, 3}];
In[19]:= Plot[f[x], {x, -3, 3}];
In[20]:= Show[%%, %];
```

See Figure 5.9; the evolute in this figure should remind you of the envelope of the normals to f — see Figure 3.6. In fact, they *are* the same, although why this is so is not at all obvious! Note also, that the points on the left side of the evolute of $y = x^2$ are the centres of the osculating circles tangent to the *right* side of the graph of $y = x^2$, and vice–versa for those on the other side.

5.2. CURVATURE

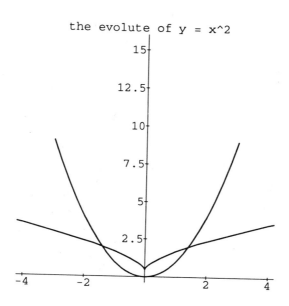

Figure 5.9: The evolute of $y = x^2$.

Example 5. It is instructive to examine the osculating circles on different sides of an inflection point of a curve. As an example, consider the inflection point $(0,0)$ on the curve of $y = \sin x$. Here are the sequence of inputs; the final output is in Figure 5.10:

```
In[21]:= viewosculatingcircle[Sin][-0.5];
In[22]:= viewosculatingcircle[Sin][0.5];
In[23]:= Show[%%, %];
```

Note that the osculating curves are on different sides of the graph, since the concavity of $y = \sin x$ changes as the curve passes through $(0,0)$. There is also an animation file, *oscinfl.ani*, which shows osculating circles moving from left to right along the curve $y = \sin x$ and passing the inflection point at $(0,0)$. You should run through it, if your work station supports animation.

Example 6. The evolute of $\sin x$ is a highly discontinuous graph. (See Figure 5.11.) You can construct a graph showing $\sin x$, its curvature function and its evolute as follows:

```
In[24]:= Plot[Sin[x], {x, -4, 4}];
In[25]:= Plot[curvature[Sin][x], {x, -4, 4}];
In[26]:= viewevolute[Sin, {-Pi, Pi}];
```

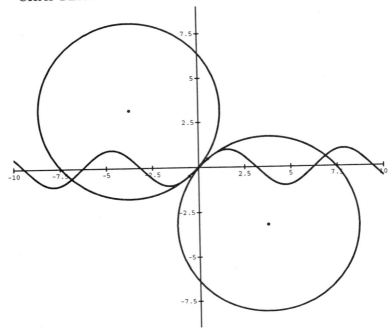

Figure 5.10: Osculating circles on different sides of an inflection point.

In[27] := Show[%%%, %%, %];

As a visual exercise, try to picture which parts of the graph of $y = \sin x$ have their centres on which part of its evolute. Near which points of $y = \sin x$ do the corresponding points of its evolute go off to infinity? Additionally, compare the evolute in Figure 5.11 with the envelope of normals to $y = \sin x$ — see Figure 3.6. Again, they are the same.

Example 7. The function **survey** constructs a graph containing various curvature details of a given function at a given point.

In[28] := ?survey
survey[f, {a, b}, s] plots f, its curvature function and its evolute for x in the interval {a, b}. In addition, it includes the osculating circle tangent to f at x = s, plus short normal and tangent lines to f at x = s.

In[28] := survey[f, {-3, 3}, 0];
In[29] := survey[f, {-3, 3}, 0.5];

Warning: **survey** may not produce very useful graphical information if the curvature and evolute of the function under consideration vary wildly — the graph will be

5.2. CURVATURE

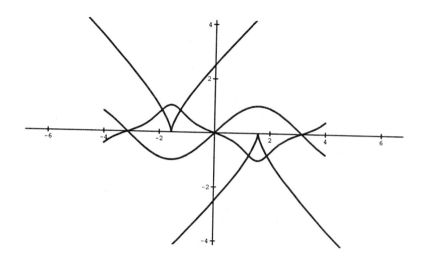

Figure 5.11: Graphs of $y = \sin x$, its curvature function, and its evolute.

cluttered with lines. You may get more information in some cases simply by creating your own combination of plots.

FUNCTIONS CONTAINED IN THE PACKAGE Chap5.m		
curvature	radiusofcurvature	centreofcurvature
viewcurvature	viewevolute	viewosculatingcircle
survey	d	cubicplot

Exercises:

1. For f as in Example 1 of Section 5.1 use `viewcurvature` to plot f and its curvature function on the same graph. Where does the maximum curvature occur?

2. Repeat Exercise 1 for the function g of Example 2 in Section 5.1.

3. Repeat Exercise 1 for the function h of Example 3 in Section 5.2. What is happening to the curvature of h near the origin? You may wish to plot `curvature[h][x]` in a small interval around the origin, using the `Plot` option `PlotPoints -> 50` to get a more detailed graph.

148 CHAPTER 5. PROPERTIES OF DIFFERENTIABLE FUNCTIONS

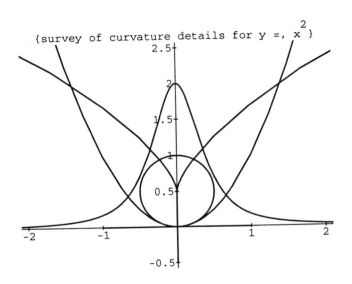

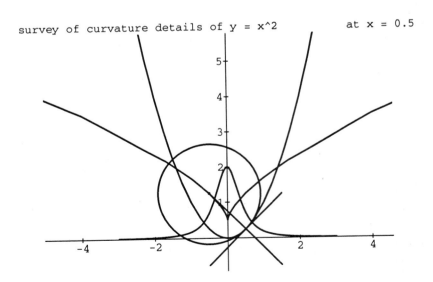

Figure 5.12: Curvature details at two differnent points of $y = x^2$.

5.2. CURVATURE

4. Repeat Exercises 1, 2, and 3 for the functions f, g, and h of Exercises 5, 6, and 7 of Section 5.1.

5. What is the maximum curvature of $y = e^x$? At what value of x does it occur?

6. Apply **survey** to the function $\sin x$ on the interval $[-4, 4]$ at the point $x = \pi/2$.

7. Consider the ellipse with equation

$$\frac{x^2}{16} + \frac{y^2}{9} = 1.$$

   (a) Graph the evolute of the top half of this ellipse on the interval $[-4, 4]$, and then graph the evolute of the bottom half on the same interval. Combine the two graphs with **Show** to view the complete evolute of the entire ellipse.

   (b) Use **ParametricPlot** to plot the ellipse; combine this graph with the one from part a) using **Show**.

   (c) For the top half of the ellipse, use the defined function **normalfamily**, which is in **Chap3.m**, to find the envelope of normals to the ellipse. Repeat for the bottom half of the ellipse, and then combine your two graphs into one with **Show**. Compare this graph with the evolute of part b).

8. This exercise will use *Mathematica* to find the co-ordinates of the centre of curvature — we shall prove the formula for **centreofcurvature[f][a]**. Let (x, y) be the centre of the osculating circle of f at $x = a$.

   (a) Explain why (x, y) and $(a, f(a))$ must satisfy the system of equations:

$$\begin{cases} (x-a)^2 + (y - f(a))^2 = r^2 \\ y + \frac{x-a}{f'(a)} = f(a) \end{cases}$$

   where r is the radius of curvature of the osculating circle of f at $x = a$.

   (b) Use *Mathematica* to solve the above system of equations for x and y. Simplify *Mathematica*'s answer; if absolute values remain in the answer, simplify it even further with a suitable replacement.

   (c) Decide which of the two ordered pairs (which form the solution set of the above system of equations) is the centre of curvature — it depends on the sign of $f''(a)$.

   (d) Compare your choice with the formula for **centreofcurvature[f][a]**.

9. For the top half of the ellipse in Exercise 7, use **centreofcurvature** to find the co-ordinates of the centre of curvature, in terms of x. Can you obtain an

equation for the evolute from these co–ordinates? (Hint: Let the centre of curvature be (X, Y) and consider the expression

$$(4X)^{2/3} + (3Y)^{2/3}.$$

Simplify it — this will involve some algebra on your part as *Mathematica* will not "go all the way.")

10. Once you have simplified the equation from Exercise 9, find the family of tangents to this curve. Compare with the family of normals to the ellipse of Exercise 7.

11. Examples 4 and 6, plus the previous exercise, suggest that the envelope of the normals to f is the evolute of f. Try and prove this as follows: first, show that every normal to f is a tangent of its evolute; secondly, show that every tangent to the evolute is a normal of f. You will need to use the formula for the co–ordinates of the centre of curvature, and you need to know that the derivative along a parametric curve $(x(t), y(t))$ is given by

$$\frac{dy}{dx} = \frac{\frac{dy}{dt}}{\frac{dx}{dt}}.$$

You may also wish to use the functions `tangent` and `normal` from `Chap3.m`.

Chapter 6

Antiderivatives

What's done cannot be undone. —W.Shakespeare, Macbeth, 1606

In general the problem of finding a function from its derivative is difficult, though not quite as bad as Lady Macbeth's situation! We can approach it systematically and display graphically the fact that a given derivative defines a family of functions. The reason we need to consider such procedures is that, in geometry as well as in the applied sciences, many important problems arise in which we seek a function about which only information on its derivative is available at the outset. What makes calculus so beautifully symmetric is the fact that there is a *limiting summation* process (integration via Riemann sums) which turns out to provide an inverse to the *limiting difference* process that is differentiation.

6.1 Antidifferentiation

If $F'(x) = f(x)$, then F is called an *antiderivative* of f and

(6.1) $$\int f(x)dx = F(x) + c$$

is called the *indefinite integral* of f. The number c is an arbitrary constant of integration; it does not effect the derivative of the right side of (6.1) since

$$\frac{d(F(x)+c)}{dx} = \frac{dF(x)}{dx} + \frac{dc}{dx} = F'(x) + 0 = F'(x) = f(x).$$

The indefinite integral of f represents the most general antiderivative of f in the sense that any two antiderivatives of f must differ by a constant. (This is a consequence of the Mean Value Theorem, see p 172, Edwards and Penney[3].) **Integrate** is *Mathematica's* built–in function for finding antiderivatives. Note, however, that it does not include arbitrary constants of integration:

```
In[1]:= Integrate[x^2, x]
           3
          x
Out[1]= --
         3
```

Example 1. Integrate produces general formulas whenever possible, but does not specify restrictions. Consider the integration formula

$$\int x^n dx = \frac{x^{n+1}}{n+1} \text{ for } n \neq -1.$$

```
In[2]:= Integrate[x^n, x]
         1 + n
        x
Out[2]= ------
         1 + n
```

Note that no restriction on n is given. However, if the special case $n = -1$ is input directly, *Mathematica* responds with a different answer:

```
In[3]:= Integrate[1/x, x]
Out[3]= Log[x]
```

Example 2. In a sense integration and differentiation are inverse operations: differentiating the antiderivative of f should return f:

```
In[4]:= Integrate[x^2 Log[x], x]
          3       3
         -x      x  Log[x]
Out[4]= --- + ---------
          9        3
In[5]:= D[%, x]
          2
Out[5]= x  Log[x]
```

In some cases simplication is required before the derivative of the antiderivative of f matches f:

```
In[6]:= Integrate[x (2 + x)^(-4), x]
             2              1
Out[6]= ---------- - ---------
             3              2
```

6.1. ANTIDIFFERENTIATION

```
              3 (2 + x)    2 (2 + x)
In[7]:= Simplify[%]
           -(2 + 3 x)
Out[7]= ----------
                3
           6 (2 + x)
In[8]:= D[%, x]
             -1          2 + 3 x
Out[8]= ---------- + ----------
                3              4
           2 (2 + x)    2 (2 + x)
In[9]:= Simplify[%]
             x
Out[9]= --------
              4
         (2 + x)
```

Example 3. **Integrate** can handle antiderivatives involving unspecified functions and the chain rule, as follows:

```
In[10]:= Integrate[Cos[f[x]] f'[x], x]
Out[10]= Sin[f[x]]
```

Such an integral would normally be handled — with paper and pencil — by making a simple substituion: let $u = f(x)$. Then $du = f'(x)dx$ and

$$\int \cos(f(x))f'(x)dx = \int \cos u\, du,$$

from which *Mathematica's* result immediately follows.

A second, slightly more complicated example, is solved just as easily:

```
In[11]:= Integrate[Cos[f[x]]^3 f[x]^2 f'[x], x]
                3
         Sin[f[x] ]
Out[11]= ----------
              3
```

Example 4. Not all functions have elementary antiderivatives! This means that, even with a computer, you will not be able to find antiderivatives for certain functions. When *Mathematica* runs up against such a function it returns your input unchanged. Consider $\int \cos(\cos x)dx$:

```
In[12]:= Integrate[Cos[Cos[x]], x]
Out[12]= Integrate[Cos[Cos[x]], x]
```

In some cases, even though no elementary antiderivative exists, *Mathematica* will respond with an answer in terms of functions you may not be familiar with. For example:

```
In[13]:= Integrate[Exp[-x^2], x]
         Sqrt[Pi] Erf[x]
Out[13]= ----------------
                2
```

In this case, *Mathematica's* answer involves the *error function*, Erf. You can query *Mathematica* about its answer:

```
In[14]:= ?Erf
Erf[z] gives the error function erf(z). Erf[z0, z1] gives the
generalized error function erf(z1) - erf(z0).
```

What isn't pointed out in this message, is that the error function is actually defined in terms of integrals. There are many other special functions like this, and *Mathematica* will often give antiderivatives in terms of them. We shall return to the problem of how to find antiderivatives in Chapters 7 and 12. For the moment, we can simply use *Mathematica* to find most of the antiderivatives that we shall need.

Example 5. The abstract relationships between antiderivatives and derivatives are built into *Mathematica*. For example, the equations

$$\frac{d(\int f(x)dx)}{dx} = f(x)$$

and

$$\int f'(x)dx = f(x) + c$$

can be generated with *Mathematica* by inputting the left-hand side, and obtaining the right-hand side as an output:

```
In[14]:= D[Integrate[f[x], x], x]
Out[14]= f[x]
In[15]:= Integrate[D[f[x], x], x]
Out[15]= f[x]
```

As before, *Mathematica* does not supply the arbitrary constant in Out[15]. The above results apply even if $f(x)$ is a function that *Mathematica* cannot integrate, say $f(x) = \cos(\cos x)$.

Example 6. If F and G are both antiderivatives of the same function then they differ by at most a constant. As an example, consider the functions

$$F(x) = \frac{1}{1-x} \quad \text{and} \quad G(x) = \frac{x}{1-x}.$$

6.1. ANTIDIFFERENTIATION

```
In[16]:= F[x_] := 1/(1 - x); G[x_] := x/(1 - x)
In[17]:= D[{F[x], G[x]}, x]
                 -2      1         x
Out[17]= {(1 - x)   , ----- + --------}
                      1 - x          2
                               (1 - x)
In[18]:= Simplify[%]
                  -2           -2
Out[18]= {(-1 + x)  , (-1 + x)   }
```

Thus $F' = G'$, and so there should be a constant c such that

$$F(x) = G(x) + c.$$

Plotting the graphs of F and G is one possible way to see what c is, but the graphs are so close together it is hard to evaluate c by inspection. It is better to simplify $F(x) - G(x)$ algebraically:

```
In[19]:= Plot[{F[x], G[x]}, {x, -2, 3}];
In[20]:= F[x] - G[x]
             1       x
Out[20]= ----- - -----
         1 - x   1 - x
In[21]:= Simplify[%]
Out[21]= 1
```

Thus $F(x) = G(x) + 1$.

Example 7. In the previous example each function was not defined at $x = 1$, nevertheless $F(x) = G(x) + 1$ for all values of $x \neq 1$. In other cases, when a discontinuity is involved things may get a little trickier. Consider the case of the function

$$f(x) = \arctan x + \arctan 1/x,$$

which has a discontinuity at $x = 0$.

```
In[22]:= f[x_] := ArcTan[x] + ArcTan[1/x]
In[23]:= f'[x]
                    1            1
Out[23]= -(-------------) + ------
                 -2   2          2
            (1 + x  ) x      1 + x
In[24]:= Simplify[%]
Out[24]= 0
```

Since $f'(x) = 0$, for all $x \neq 0$, one would expect f to be constant. We can plot f to see:

```
In[25]:= Plot[f[x], {x, -3, 3}];
                           1
Power::infy: Infinite expression -- encountered.
                           0.
Plot::notnum: f[x] does not evaluate to a real number at x=0..
```

Note that f is constant on each side of $x = 0$, but it is a *different* constant on each side. The constants are $-\pi/2$ and $\pi/2$, as you can check by evaluating f at $x = -1$ and $x = 1$, say.

```
In[26]:= f[-1]
         -Pi
Out[26]= ---
          2
In[27]:= f[1]
          Pi
Out[27]= --
          2
```

Can you explain why the constants are $\pm\pi/2$?

Example 8. Since any two antiderivatives of f will differ by at most a constant, the family of antiderivatives of f must consist of a family of curves of the same shape, but shifted vertically from each other. There is a defined function, antiderivativefamily, in the package Chap6.m that will allow you to see this. For the following examples, your graphical output should look something like the graphs in Figures 6.1 and 6.2

```
In[28]:= << Chap6.m (* Out[28] will be a usage message *)
In[29]:= ?antiderivativefamily
antiderivativefamily[f, {a, b, k}] plots F[x] + c, and f[x], on the
interval {a, b} where F[x] = Integrate[f[x], x] and values
  of c differ by k. The default value for k is 1.
In[29]:= antiderivativefamily[Sin, {-Pi, Pi, 0.5}];
In[30]:= antiderivativefamily[Log, {0.1, 6, 0.5}];
In[31]:= f[x_] := x^2
In[32]:= antiderivativefamily[f, {-2, 2}];
In[33]:= antiderivativefamily[ArcTan, {-6, 6}];
```

Example 9. The acceleration due to gravity (at the earth's surface) is 32 ft/sec^2, or 9.8 m/sec^2 in metric units. Consider a particle close enough to the earth's surface

6.1. ANTIDIFFERENTIATION

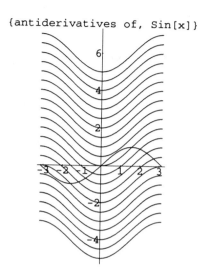

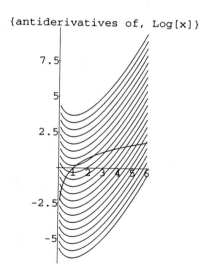

Figure 6.1: Antiderivatives of some well known functions: $\sin x$ (top); $\log x$ (bottom). A particular antiderivative must be specified by an initial condition.

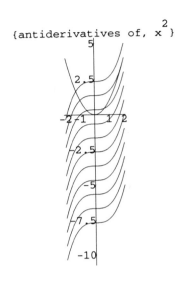

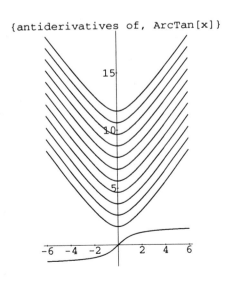

Figure 6.2: Antiderivatives of some well known functions: x^2 (top); $\arctan x$ (bottom). A particular antiderivative must be specified by an initial condition.

6.1. ANTIDIFFERENTIATION

so that the acceleration due to gravity can be considered constant. Let $s(t)$ be the position of the particle at time t. Then its velocity, $v(t)$, and acceleration, $a(t)$, are given by

$$v(t) = s'(t) \text{ and } a(t) = v'(t).$$

If we take *up* as the positive direction, then we can find v by integrating:

$$v(t) = \int a(t)dt = \int -32 dt = -32t + c.$$

To evaluate c let us assume an initial condition of $v_0 = v(0)$. Thus $c = v_0$ and

$$v(t) = -32t + v_0.$$

Similarily, by integrating v we can find s to be

$$s(t) = -16t^2 + v_0 t + s_0,$$

where we are assuming the initial condition $s_0 = s(0)$. We can define these functions in *Mathematica*, but since it does not accept subscripts, let us use *vinit* and *sinit* for the initial conditions v_0 and s_0, respectively.

```
In[34]:= s[t_] := -16 t^2 + vinit t + sinit
In[35]:= v[t_] := s'[t]
In[36]:= a[t_] := v'[t]
```

Once these functions are defined in *Mathematica* then you can use it to solve all your homework problems about rectilinear motion in the earth's gravitational field. Consider, for example, question 15, p 220, Edwards and Penney[3] —funny units again.

> A ball is thrown upward with an initial velocity of 48 ft/s from the top of a building 160 ft tall, then falls to the ground at the base of the building. How long does the ball remain aloft, and with what speed does it strike the ground?

```
In[37]:= s[t] /. {vinit -> 48, sinit -> 160}
                 2
Out[37]= 160 + 48 t - 16 t
In[38]:= Solve[% == 0, t]
Out[38]= {{t -> -2}, {t -> 5}}
In[39]:= v[5] /. {vinit -> 48, sinit -> 160}
Out[39]= -112
```

Thus the ball will stay aloft for 5 seconds, and will hit the ground with a speed of 112 ft/sec. Note that replacement values were used in the above calculations. This allows you to use the same functions, s, v and a, for the next question without having to redefine them. Question 16:

A ball is dropped from the top of a building 576 ft high. With what velocity should a second ball be thrown straight downward 3 s later in order that the two balls hit ground simultaneously?

```
In[40]:= s[t] /. {vinit -> 0, sinit -> 576}
                 2
Out[40]= 576 - 16 t
In[41]:= Solve[% == 0, t]
Out[41]= {{t -> 6}, {t -> -6}}
In[42]:= s[t] /. {sinit -> 576, t -> 3}
Out[42]= 432 + 3 vinit
In[43]:= Solve[% == -576, vinit]
Out[43]= {{vinit -> -336}}
```

Thus the second ball must be thrown downwards with an initial velocity of 336 ft/sec.

Exercises:

1. For each of the following functions f find an antiderivative F satisfying the given initial condition:

    (a) $f(x) = \dfrac{1}{\sqrt{x+1}}$; $F(2) = -1$

    (b) $f(x) = x^4 - 3x + \dfrac{3}{x^3}$; $F(1) = -1$

    (c) $f(x) = x\sqrt{1-x^2}$; $F(1) = 0$

    (d) $f(x) = \sin x + \cos 2x$; $F(\pi) = 3$

    (e) $f(x) = 40e^{0.02x}$; $F(0) = 500$

2. Let $F(x) = \sin^2 x$; let $G(x) = -\cos^2 x$.

    (a) Show that $F'(x) = G'(x)$ for all x.
    (b) Find a constant c such that $F(x) - G(x) = c$ for all x.

3. Repeat Exercise 2 for the pair of functions

$$F(x) = -\sqrt{1-x^2} + \dfrac{(1-x^2)^{3/2}}{3}$$

and

$$G(x) = -x^2\sqrt{1-x^2} - \dfrac{2}{3}(1-x^2)^{3/2},$$

for $|x| \leq 1$.

6.1. ANTIDIFFERENTIATION

4. Use **Integrate** to find
$$\int x^t dx \text{ and } \int x^t dt.$$
Note the different variables of integration; note the different answers.

5. **Integrate** is listable — you can integrate a list of functions to find many indefinite integrals at once. Try integrating all of the following at once:
$$x^2, \quad \sin x, \quad 1/x, \quad e^x$$

6. Can **Integrate** handle tables? Try integrating `Table[x^i, {i, -2, 2 }]` with respect to x.

7. Can **Integrate** handle sums?

    (a) Try integrating the sum $\sum_{i=0}^{8} x^i$.

    (b) Now try integrating the sum $\sum_{i=0}^{n} x^i$.

8. Use `antiderivativefamily` to plot some antiderivatives of the following functions:

    (a) $f(x) = x^{2/3}$ on $[-4, 4]$. Where does each antiderivative have an inflection point? Explain, in terms of f.

    (b) $g(x) = xe^{-x}$ on $[-1, 6]$. What do you notice about each antiderivative as $x \to \infty$? Explain, in terms of g.

    (c) $h(x) = \dfrac{x^2 - 1}{1 + x^2}$ on $[-10, 10]$. What do you notice about each antiderivative as $x \to \pm\infty$? Explain in terms of h.

9. Use **Integrate** to find the following indefinite integrals. If *Mathematica* does not return an answer, can you find an antiderivative?

    (a) $\int |x| dx$

    (b) $\int \dfrac{1}{1 - |x|} dx$

    (c) $\int \dfrac{\cos x}{\sqrt{\sin x}} dx$

    (d) $\int [[x]] dx$ (Hint: consider one interval $[n, n+1]$ at a time.)

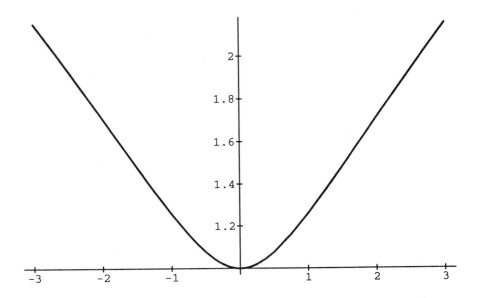

Figure 6.3: The graph of $y = (1 + x^2)^{1/3}$. From its graph we can see that its antiderivatives must be increasing functions with an inflection point at $x = 0$.

6.2 Constructing Antiderivatives

In this section we will briefly consider the problem of plotting antiderivatives for given functions. This will be particularly useful in the case where *Mathematica* cannot integrate the given function. We will not bother about finding a specific *equation* for the antiderivative; we will simply try to plot its graph. (This is what we do all the time with paper and pencil, when we use information about f' to sketch the graph of f.) Of course, we will need to have an initial condition to specify one particular antiderivative.

Example 1. As an opening example, consider the function $f(x) = (1+x^2)^{1/3}$; its graph is plotted in Figure 6.3.

```
In[1]:= f[x_] := (1 + x^2)^(1/3)
In[2]:= Plot[f[x], {x, -3, 3}];
```

It so happens that f is one of those functions that has no elementary antiderivative. We can try integrating it with *Mathematica* but we will get no useful results; nevertheless let us try *plotting* the output that we get:

```
In[3]:= Integrate[f[x], x]
```

6.2. CONSTRUCTING ANTIDERIVATIVES

```
                              2  1/3
Out[3]= Integrate[(1 + x )       , x]
In[4]:= Plot[%, {x, -3, 3}];
General::bvar: -3. is a number which cannot be used as a variable.
NIntegrate::vars: Integration range specification -3.
   is not of the form {x, xmin, xmax}.
Plot::notnum: % does not evaluate to a real number at x=-3..
General::bvar: -2.75 is a number which cannot be used as a variable.
NIntegrate::vars: Integration range specification -2.75
is not of the form {x, xmin, xmax}.
Plot::notnum: % does not evaluate to a real number at x=-2.75.
General::bvar: -2.5 is a number which cannot be used as a variable.
General::stop: Further output of General::bvar
     will be suppressed during this calculation.
NIntegrate::vars: Integration range specification -2.5
is not of the form {x, xmin, xmax}.
General::stop: Further output of NIntegrate::vars
     will be suppressed during this calculation.
Plot::notnum: % does not evaluate to a real number at x=-2.5.
General::stop: Further output of Plot::notnum
     will be suppressed during this calculation.
```

Mathematica responds with error messages and produces no graph at all. This is not surprising — something more than a naive approach is required!

Example 2. If F is an antiderivative of f on the interval $[a,b]$, then the Mean Value Theorem implies there is a number c in the interval $[a,b]$ such that

$$F(b) = F(a) + f(c)(b - a).$$

Let us consider $F(a)$ as an initial value. Then, if we knew the value of c, we would know another value of F, namely $F(b)$. Unfortunately, the Mean Value Theorem simply implies the existence of such a c; it does not state its value. However, in some special cases we can find c explicitly. Consider a general quadratic function

$$F(x) = s + tx + ux^2.$$

```
In[5]:= F[x_] := s + t x + u x^2
In[6]:= Solve[F'[c] == (F[b] - F[a])/(b - a), c]
                    a u + b u
Out[6]= {{c -> ---------}}
                       2 u
In[7]:= Simplify[%]
                   a + b
Out[7]= {{c -> -----}}
                     2
```

Thus c is the midpoint of the interval $[a, b]$. Since F' is a linear function,
$$F'(\frac{a+b}{2}) = \frac{F'(a) + F'(b)}{2}$$
as we can check with *Mathematica*:

```
In[8] := F'[(a + b)/2]
Out[8]= t + (a + b) u
In[9] := (F'[a] + F'[b])/2
        2 t + 2 a u + 2 b u
Out[9]= -------------------
                 2
In[10] := Simplify[%]  (* to see if Out[9] == Out[8] *)
Out[10]= t + a u + b u
```

Thus for quadratic F, with derivative f, on any interval $[a, b]$, we have:
$$F(b) = F(a) + \frac{f(a) + f(b)}{2}.$$

In this case, we can use the given function f and an initial condition $F[a]$ to find as many points on the antiderivative F as we wish. Unfortunately, if f is linear then we can integrate it explicitly, so finding its antiderivative, F, is no problem anyway. However, the results of Example 2 do suggest a *strategy* for plotting general antiderivatives: take many points from the graph of f, apply the results of Example 2 on each pair of consecutive points chosen on f to produce a set of data points for F, and then join these with straight lines to produce an *approximate* antiderivative of f. This procedure is illustrated in the next example, for the function
$$f(x) = e^x \sin \pi x,$$
which *can* be integrated, so we will be able to check our results.

Example 3. (This example, and the next one, can be passed over on a first reading.) The graph of f on $[0, 2]$ is shown in Figure 6.4; from it we can see that any antiderivative of f must be increasing on $[0, 1]$, decreasing on $[1, 2]$, and have two inflection points.

```
In[11] := f[x_] := Exp[x] Sin[Pi x]
In[12] := Plot[f[x], {x, 0, 2}];
```

Let us start by taking five points on the graph of f, say for x equally spaced from 0 to 2:

```
In[13] := Table[{i, f[i]}, {i, 0, 2, 1/2}] // N
Out[13]= {{0., 0.}, {0.5, 1.64872}, {1., 0.}, {1.5, -4.48169}, {2., 0.}}
```

6.2. CONSTRUCTING ANTIDERIVATIVES

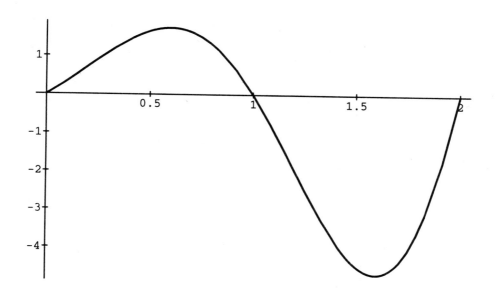

Figure 6.4: The graph of $y = e^x \sin \pi x$, on $[0, 2]$.

As an initial value, let us take $F(0) = 0$, for simplicity. Then, using the results from Example 2,
$$F(0.5) \simeq 0 + \frac{f(0) + f(0.5)}{2} = 0.82436.$$
This calculation, and subsequent ones for $F(1)$, $F(1.5)$ and $F(2)$ can be done as follows:

```
In[14]:= (0 + 1.64872)/2     (* to find F[0.5], approximately *)
Out[14]= 0.82436
In[15]:= % + (1.64872 + 0)/2   (* F[1], approximately *)
Out[15]= 1.64872
In[16]:= % + (0 - 4.48169)/2   (* F[1.5], approximately *)
Out[16]= -0.592125
In[17]:= % + (- 4.48169 + 0)/2   (* F[2], approximately *)
Out[17]= -2.83297
```

Plotting these points, and joining them with straight lines, will give us an approximate antiderivative of f:

```
In[18]:= ListPlot[{{0, 0}, {0.5, Out[14]}, {1, Out[15]},
{1.5, Out[16]}, {2, Out[17]}}, PlotJoined -> True];
```

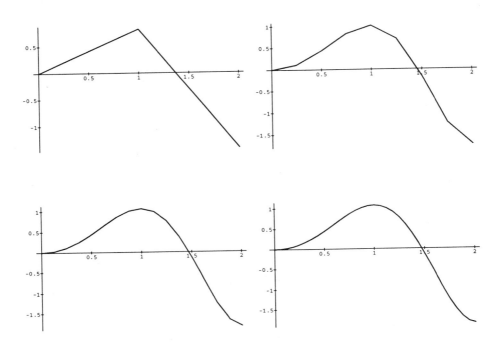

Figure 6.5: Constructing an approximate antiderivative of $f(x) = e^x \sin \pi x$. Five points from f are used in the top left, 9 points are used in the top right, 17 points are used in the bottom left, 33 are used in the bottom right.

This output is the top left part of Figure 6.5. By picking more points from f we should be able to smooth this graph out. To help with calculations of this nature, two (rather technical) functions have been defined in Chap6.m, anti and antiderivative. We can use these to extend the above calculations to include more and more points from f. The calculations are below, and the resulting approximations to an antiderivative of f are shown in the remaining parts of Figure 6.5.

```
In[19]:= << Chap6.m  (* Out[19] will be a usage statement *)
In[20]:= ?anti
anti
anti/: anti[tableofvalues_][1] = 0

anti/: anti[tableofvalues_][n_] :=

 >      anti[tableofvalues][n] =
```

6.2. CONSTRUCTING ANTIDERIVATIVES

```
>           anti[tableofvalues][n - 1] +
>             ((First[Transpose[tableofvalues]][[n]] -
>                First[Transpose[tableofvalues]][[n - 1]])
>               (Last[Transpose[tableofvalues]][[n]] +
>                Last[Transpose[tableofvalues]][[n - 1]])) / 2
In[20]:= ?antiderivative
antiderivative
antiderivative/:

>     antiderivative[tableofvalues_, k_] :=
>       Block[{i}, Table[{First[Transpose[tableofvalues]][[i]],
>         anti[tableofvalues][i] + k}, {i, Length[tableofvalues]}]]
In[20]:= antiderivative[N[Table[{i, f[i]}, {i, 0, 2, 0.25}]], 0]
Out[20]= {{0., 0}, {0.25, 0.113493},
 {0.5, 0.433076}, {0.75, 0.826284},
  {1., 1.0134}, {1.25, 0.704897},
 {1.5, -0.16382}, {1.75, -1.23267},
  {2., -1.74131}}
In[21]:= ListPlot[%, PlotJoined -> True];
In[22]:= antiderivative[N[Table[{i, f[i]}, {i, 0, 2, 0.125}]], 0]
Out[22]= {{0., 0}, {0.125, 0.0271023},
 {0.25, 0.110951}, {0.375, 0.251712},
  {0.5, 0.438772}, {0.625, 0.649694},
  {0.75, 0.851131}, {0.875, 1.00207},
  {1., 1.05944}, {1.125, 0.985769},
  {1.25, 0.757845}, {1.375, 0.375216},
  {1.5, -0.133266}, {1.625, -0.706612},
  {1.75, -1.25417}, {1.875, -1.66445}, {2., -1.82042}}

In[23]:= ListPlot[%, PlotJoined -> True];
In[24]:= antiderivative[N[Table[{i, f[i]}, {i, 0, 2, 1/16}]], 0]
Out[24]= {{0., 0}, {0.0625, 0.00648977}, {0.125, 0.0265307},

>      {0.1875, 0.0610239}, {0.25, 0.110339}, {0.3125, 0.174228},
```

>   {0.375, 0.25175}, {0.4375, 0.341228}, {0.5, 0.440222},

>   {0.5625, 0.545536}, {0.625, 0.653266}, {0.6875, 0.758879},

>   {0.75, 0.857333}, {0.8125, 0.943237}, {0.875, 1.01105},

>   {0.9375, 1.05531}, {1., 1.07087}, {1.0625, 1.05323},
{1.125, 0.998756},
>   {1.1875, 0.904994}, {1.25, 0.770941}, {1.3125, 0.597274},

>   {1.375, 0.386546}, {1.4375, 0.143319}, {1.5, -0.125773},

>   {1.5625, -0.412047}, {1.625, -0.704888}, {1.6875, -0.991973},

>   {1.75, -1.2596}, {1.8125, -1.49311}, {1.875, -1.67744},

>   {1.9375, -1.79774}, {2., -1.84006}}
In[25]:=ListPlot[%,PlotJoined ->True];

The next, and final calculation, will use 65 points from f. The resulting approximate antiderivative is shown in the top of Figure 6.6; the bottom of Figure 6.6 shows an actual antiderivative of f, for comparison.

In[26]:= antiderivative[N[Table[{i, f[i]}, {i, 0, 2, 1/32}]], 0]
Out[26]= {{0., 0}, {0.03125, 0.00158013}, {0.0625, 0.00640515},

>   {0.09375, 0.0146315}, {0.125, 0.0263886}, {0.15625, 0.0417754},

>   {0.1875, 0.0608577}, {0.21875, 0.0836649}, {0.25, 0.110188},

>   {0.28125, 0.140375}, {0.3125, 0.174134}, {0.34375, 0.211325},

>   {0.375, 0.251761}, {0.40625, 0.295211}, {0.4375, 0.341392},

>   {0.46875, 0.389976}, {0.5, 0.440586}, {0.53125, 0.492798},

>   {0.5625, 0.546145}, {0.59375, 0.600116}, {0.625, 0.65416},

>   {0.65625, 0.707691}, {0.6875, 0.76009}, {0.71875, 0.81071},

>   {0.75, 0.858883}, {0.78125, 0.903923}, {0.8125, 0.945136},

>   {0.84375, 0.981824}, {0.875, 1.01329}, {0.90625, 1.03886},

6.2. CONSTRUCTING ANTIDERIVATIVES

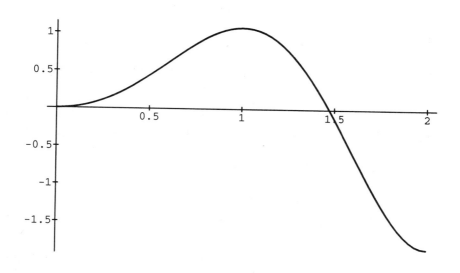

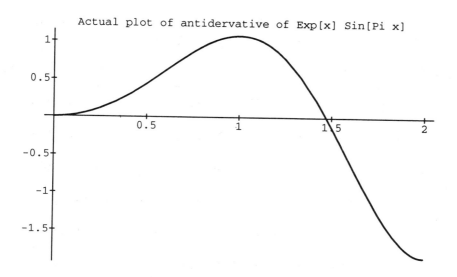

Figure 6.6: Approximate (top) and actual (bottom) antiderivatives of $f(x) = e^x \sin \pi x$. The approximation is based on 65 data points from the graph of f.

> {0.9375, 1.05787}, {0.96875, 1.06969}, {1., 1.07373},

> {1.03125, 1.06943}, {1.0625, 1.05632}, {1.09375, 1.03395},

> {1.125, 1.002}, {1.15625, 0.96017}, {1.1875, 0.908299},

> {1.21875, 0.846303}, {1.25, 0.774206}, {1.28125, 0.692148},

> {1.3125, 0.600382}, {1.34375, 0.499288}, {1.375, 0.389369},

> {1.40625, 0.271261}, {1.4375, 0.145727}, {1.46875, 0.0136625},

> {1.5, -0.123909}, {1.53125, -0.265837}, {1.5625, -0.410849},

> {1.59375, -0.557556}, {1.625, -0.704464}, {1.65625, -0.849976},

> {1.6875, -0.992411}, {1.71875, -1.13001}, {1.75, -1.26096},

> {1.78125, -1.38339}, {1.8125, -1.49542}, {1.84375, -1.59515},

> {1.875, -1.68069}, {1.90625, -1.7502}, {1.9375, -1.80187},

> {1.96875, -1.834}, {2., -1.84497}}

In[27]:= ListPlot[%, PlotJoined -> True];

At this stage, having taken 65 points from the graph of f our approximate antiderivative, top of Figure 6.6, is a smooth curve. We can see how accurate it is by integrating f, and then plotting an actual antiderivative, which is shown in the bottom of Figure 6.6.

```
In[28]:= Integrate[f[x], x]
              x                  x
           E  Pi Cos[Pi x]     E  Sin[Pi x]
Out[28]= -(---------------) + -------------
                      2                2
                1 + Pi           1 + Pi
In[29]:= Plot[%, {x, 0, 2}];
In[30]:= Show[%, Out[27]];
```

The two graphs are very similar, but they do not have the same initial value at $x = 0$. To adjust this discrepancy, we can add $\pi/(1 + \pi^2)$ to the antiderivative of Out[28]. Then we see that the agreement is very good indeed.

6.2. CONSTRUCTING ANTIDERIVATIVES

```
In[31]:= Plot[Out[30] + Pi/(1 + Pi^2), {x, 0, 2}] ;
In[32]:= Show[%, Out[29]];
```

(There is an animation file, *anti.ani* which you can use to see the above antiderivative taking shape as more and more points are used.)

It seems as if we have developed a fairly good way to approximate the antiderivative of a function, but it relies on producing a lot of data points. Is there an easy way to get these? In terms of *Mathematica* there is a *very* easy way. To plot a graph *Mathematica* actually plots only a sample of points (judiciously chosen, of course), and joins them by small straight lines. (Mostly these little straight lines that form a plot are hardly noticeable — they seem to form a nice, smooth curve.)

Example 4. We can access these points that *Mathematica* uses to plot a graph by using the function `InputForm`.

```
In[33]:= InputForm[Out[12]]
Out[33]//InputForm=
  Graphics[{{Line[{{0., 0.}, {0.125, 0.4336371393988049},
  {0.25, 0.907943079355784}, {0.3125, 1.136484213028186},
  {0.375, 1.3442367879454}, {0.4375, 1.519069958742897},
  {0.46875, 1.590300665082535}, {0.5, 1.648721270700128},
  {0.515625, 1.672667621564038}, {0.53125, 1.692866245993745},
  {0.546875, 1.709143742775439}, {0.5546875, 1.715759248107803},
  {0.5625, 1.721331773849802}, {0.5703125, 1.725841169665553},
  {0.57421875, 1.727691003935179}, {0.578125, 1.729267678707471},
  {0.58203125, 1.730568789042860}, {0.5859375, 1.731591957022255},
  {0.58984375, 1.732334832354905}, {0.59375, 1.73279509298642},
  {0.59765625, 1.732970445706866}, {0.6015625, 1.732858626758828},
  {0.60546875, 1.732457402445325}, {0.609375, 1.731764569737506},
  {0.61328125, 1.730777956881991}, {0.6171875, 1.729495424007776},
  {0.625, 1.726034201768583}, {0.6328125, 1.721364444723682},
  {0.640625, 1.715470243424159}, {0.65625, 1.699947730107002},
  {0.671875, 1.679351093800377}, {0.6875, 1.653574772805133},
  {0.71875, 1.586114464813877}, {0.75, 1.496945067518856},
  {0.8125, 1.251996846849822}, {0.875, 0.918009831311143},
  {1., 3.330298782142777*10^-16}, {1.125, -1.178747956172733},
  {1.25, -2.468045173887977}, {1.3125, -3.089284384605095},
  {1.375, -3.654014433818134}, {1.4375, -4.129260265008849},
  {1.46875, -4.32288539968019}, {1.5, -4.481689070338065},
  {1.515625, -4.546782000749336}, {1.53125, -4.601687554496479},
  {1.546875, -4.645934378210957}, {1.5546875, -4.663917186141997},
  {1.5625, -4.679064881605091}, {1.5703125, -4.691322690308378},
  {1.57421875, -4.696351061189163}, {1.578125, -4.700636907572074},
  {1.58203125, -4.704173692153584}, {1.5859375, -4.706954951079434},
```

{1.58984375, -4.708974295596985}, {1.59375, -4.710225413707986},
{1.59765625, -4.710702071821549}, {1.6015625, -4.710398116407018},
{1.60546875, -4.709307475646487}, {1.609375, -4.707424161086661},
{1.61328125, -4.704742269289789}, {1.6171875, -4.70125598348341},
{1.625, -4.691847405966353}, {1.6328125, -4.679153690247788},
{1.640625, -4.663131589962107}, {1.65625, -4.620937024080065},
{1.671875, -4.564949561880387}, {1.6875, -4.494882256914489},
{1.71875, -4.311506127559606}, {1.75, -4.069118575237906},
{1.8125, -3.403280278079893}, {1.875, -2.495409442799837},
{2., -(1.8105381325676*10^-15)}, {2.125, 3.204169149597582},
{2.25, 6.708842347995733}, {2.3125, 8.39754560561432},
{2.46875, 11.75082082846158}, {2.5, 12.18249396070347},
{2.515625, 12.35943489060158}, {2.53125, 12.50868365963392},
{2.546875, 12.62895899650402}, {2.5546875, 12.67784133652763},
{2.5625, 12.71901704184799}, {2.5703125, 12.75233722050286},
{2.57421875, 12.76600574969485}, {2.578125, 12.77765588803709},
{2.58203125, 12.78726986529618}, {2.5859375, 12.79483011089456},
{2.58984375, 12.80031925840202}, {2.59375, 12.8037201500284},
2.59765625, 12.80501584111689}, {2.6015625, 12.80418960463691},
{2.60546875, 12.80122493567619}, {2.609375, 12.79610555593093},
{2.61328125, 12.7888154181936}, {2.6171875, 12.77933871083731},
{2.625, 12.75376354554104}, {2.6328125, 12.7192584487679},
{2.640625, 12.67570586470733}, {2.65625, 12.56100914301046},
{2.671875, 12.40881944189154}, {2.6875, 12.21835676003364},
{2.71875, 11.71988875983511}, {2.75, 11.06101108091437},
{2.8125, 9.25107493705762}, {2.875, 6.783226142927911},
{3., 7.382329358236023*10^-15}}]}},
  {PlotRange -> Automatic, AspectRatio -> GoldenRatio^(-1),
    DisplayFunction ->
     ((Display["D:\\tempfile", #1];
        Run["msdosps -monitor 1 -device vgahi D:\\tempfile"];
        Run["del D:\\tempfile"]; #1) &), PlotColor -> Automatic,
    Axes -> Automatic, PlotLabel -> None, AxesLabel -> None,
    Ticks -> Automatic, Framed -> False, Prolog -> {}, Epilog -> {},
 AxesStyle ->{},Background ->Automatic,DefaultColor ->Automatic}]

The bulk of this information is the list of points that *Mathematica* uses to plot the function $f(x) = e^x \sin \pi x$. We can extract this table of values, and use it as a basis for approximating the antiderivative of f. There is a defined function, **viewantiderivative**, that will do all this for you automatically.

In[34]:= ?viewantiderivative
viewantiderivative[f,{a,b},initialvalue] plots the graph of an

6.2. CONSTRUCTING ANTIDERIVATIVES

approximate antiderivative F of f on the interval {a, b} with initial condition F[a]=initialvalue. The approximation applies the Mean Value Theorem to Mathematica InputForm of the Plot of f on {a,b}.

Example 5. Having gone to great lengths to explain what is behind this construction of approximate antiderivatives, we can now simply use **viewantiderivative** and have *all* the work done by *Mathematica*!

In[34]:= viewantiderivative[f, {0, 2}, 0];

Finally, we can produce an antiderivative for the function

$$f(x) = (1 + x^2)^{1/3}$$

with which we started this section:

In[35]:= f[x_] := (1 + x^2)^(1/3)
In[36]:= viewantiderivative[f, {-3, 3}, 0];

For all its apparent accuracy, this method is still only an approximation — perhaps suitable for working with a computer, but an approximation nonetheless. Can we find the precise, exact antiderivative? The answer is yes, as we shall see in the next chapter — but our approach will be somewhat different.

MATHEMATICA FUNCTIONS INTRODUCED IN THIS SECTION
InputForm

FUNCTIONS CONTAINED IN THE PACKAGE Chap6.m		
anti	antiderivative	viewantiderivative
antiderivativefamily		

Exercises:

1. Use **viewantiderivative** to plot the antiderivative of $f(x) = \cos x$ on the interval $[0, \pi]$, with initial condition $F(0) = 0$. What is the exact antiderivative? Compare graphs of the approximation produced by **viewantiderivative** and the exact antiderivative.

2. Repeat Exercise 1 for the function $f(x) = x^2$ on $[0, 4]$.

3. Repeat Exercise 1 for the function $f(x) = e^{-x^2}$ on the interval $[-2, 2]$. Recall that *Mathematica* supplies the function `Erf` as an antiderivative of f. (Don't worry about matching initial conditions; your two plots should have the same shape and scale.)

4. By using different initial conditions you can plot many antiderivatives of a given function. Plot antiderivatives of $f(x) = \cos x$ on the interval $[0, \pi]$, with initial conditions $F(0) = 0, -1, +1, -2,$ and 2, in turn. Compare them all on one graph.

5. Use `viewantiderivative` to plot an antiderivative, F, of the given function with the given initial conditions:

    (a) $|x|$ on $[-3, 3]$; $F(-3) = -2$.
    (b) $\sqrt{2 - \sin^2 x}$ on $[-2, 2]$; $F(-2) = 1$
    (c) *Mathematica* cannot find anitderivatives for either of the above two functions. Which one can you evaluate on your own?

6. Use `viewantiderivative` to plot the antiderivative of $f(x) = \cos(\cos x)$ on the interval $[-\pi, \pi]$, with initial condition $F(-\pi) = 0$. Recall that f is a function that *Mathematica* cannot integrate.

7. Use `viewantiderivative` to plot the antiderivative of
$$f(x) = \frac{x^2 + 15x + 50}{(1 + x^2)^{1/3}},$$
on the interval $[-15, 15]$ with initial condition $F(-15) = 0$. Compare with your answer of Exercise 9, Section 5.1.

8. To access the table of values in the `InputForm` of a plot made by *Mathematica* we can use the function `First`, which returns the first element of its argument. If you look through the input form of the plot in Example 4, you will notice that, starting from the left, we have to "peel off" four levels of brackets before we get to the list which contains the table of values. Thus we can apply `First` four times, in succession, to the input form of a plot, to obtain the table of values used by *Mathematica* to construct the plot.

    (a) To do this for the graph of sine on the interval $[0, \pi]$, plot the graph first, enter `InputForm[%]`, and then enter `Nest[First, %, 4]`. The output should be a list of points.
    (b) Use `ListPlot` to plot the points produced in part a). What do you notice about the number of points in relation to the curvature of the curve?
    (c) Now use the option `PlotJoined -> True` on the plot of part b); this should produce the original sine curve.

Chapter 7

The Definite Integral

The properties which distinguish space from other conceivable triply-extended magnitudes are only to be deduced from experience. —G.F.B.Riemann, On the hypotheses that lie at the foundations of geometry, 1854

7.1 Partitions and Riemann Sums

The calculation of Riemann sums is one calculus topic that is ideally suited to being done by computer. Indeed, it is the difficulty of calculating Riemann sums in any but the most simple cases that probably makes this topic seem so difficult to the beginning student. In the package for this Chapter, Chap7.m, we have included many functions to make the handling of partitions and Riemann sums easy and painless. Let us remind you that any set of points

$$P = \{x_0, x_1, x_2, \ldots, x_n\}$$

such that

$$a = x_0 < x_1 < x_2 < \cdots < x_n = b$$

is called a *partition* of the interval $[a, b]$.

If the points are equally spaced so that $x_i - x_{i-1} = \dfrac{b-a}{n}$, then P is called a *regular partition*.

Example 1. The package for Chapter 7 includes three defined functions to generate partitions of a given interval: `regularpartition`, `randompartition`, and `refine`.

```
In[1] := << Chap7.m
```
(* Out[1] will be a usage statement *)
```
In[2] := ?regularpartition
regularpartition[{a, b}, n] partitions the interval {a, b} into n
equally spaced subintervals. The partition is set equal to P.
```

Thus it is an easy matter to generate a regular partition of the interval [2, 6] into 8 subintervals, say:

```
In[2] := regularpartition[{2, 6}, 8]
Out[2] = {2., 2.5, 3., 3.5, 4., 4.5, 5., 5.5, 6.}
```

Any partition generated by a function in the package for this chapter is always set equal to P for future reference. This means, of course, that only the most recent partition is stored as P. So for the example above:

```
In[3] := P
Out[3] = {2., 2.5, 3., 3.5, 4., 4.5, 5., 5.5, 6.}
In[4] := norm[P]
Out[4] = 0.5
```

Note also that the defined function `norm` gives the norm of any partition P — that is, the length of the longest subinterval $[x_{i-1}, x_i]$.

In addition to regular partitions there is a defined function which generates random partitions of a given interval. (Of course, each time you use it you will probably get a different partition. Don't expect to get the same output as the following.)

```
In[5] := randompartition[{2, 6}, 8]
Out[5] = {2., 2.166, 2.41454, 2.88798, 4.76321, 5.34355, 5.4032,
5.71048, 6.}
In[6] := norm[P]
Out[6] = 1.87523
In[7] := randompartition[{2, 6}, 8]
Out[7] = {2., 2.78625, 3.1048, 3.53696, 4.75394, 5.31919, 5.33773,
5.95515, 6.}
In[8] := norm[P]
Out[8] = 1.21698
```

Finally, there is also a function `refine` which can be used to randomly insert up to n additional points into a partition P:

7.1. PARTITIONS AND RIEMANN SUMS

```
In[9]:= refine[P, 6]
Out[9]= {2., 2.3321, 2.50487, 2.78625, 3.1048, 3.29658, 3.53696,
 3.58895, 4.20811, 4.39804, 4.75394, 5.31919, 5.33773, 5.95515,6.}
In[10]:= norm[P]
Out[10]= 0.619156
```

Note that everytime you refine a partition P, the new partition is again set equal to P — P is always equal to the most recent partition you have created. Note also that everytime you refine a partition the norm will be no bigger than before; indeed it will probably be less.

Once you have a partition P you can pick any point

$$x_i^* \text{ in the subinterval } [x_{i-1}, x_i]$$

and form a sum of the following type:

$$\sum_{i=1}^{n} f(x_i^*)(x_i - x_{i-1}).$$

Such a sum is called a *Riemann sum* for f on the interval $[a, b]$. As described in Section 5.3 of Edwards and Penney[3], the *definite integral* of f on the interval $[a, b]$ is defined as

$$\int_a^b f(x)dx = \lim_{norm P \to 0} \sum_{i=1}^{n} f(x_i^*)(x_i - x_{i-1}).$$

This limit always exists if f is continuous on $[a, b]$.

Example 2. If P is a regular partition of $[a, b]$ into n subintervals, and each x_i^* is chosen be the right end point of $[x_{i-1}, x_i]$, then we obtain the following approximation formula:

(7.1) $$\int_a^b f(x)dx \simeq \frac{b-a}{n}[f(x_1) + f(x_2) + \cdots + f(x_n)]$$

This formula is defined in the package for Chapter 7 as `riemann`. Along with it is defined the error term `er`:

```
In[11]:= ?riemann
riemann[f, {a, b}, n] calculates the numerical value of the Riemann
sum obtained by dividing the interval {a, b} into n equal subintervals
and evaluating f at the right hand end point of each subinterval.
In[11]:= ?er
er[f, {a, b}, n] computes the absolute value of the difference
between the value of the definite integral of f on the interval
{a, b} and the value of the Riemann sum approximation
riemann[f, {a, b}, n].
```

Here are some sample calculations using `riemann` to approximate

$$\int_0^\pi \sin x\, dx.$$

Note that increasing n results in better approximations:

```
In[11]:= riemann[Sin, {0, Pi}, 8]
Out[11]= 1.97423
In[12]:= er[Sin, {0, Pi}, 8]
Out[12]= 0.0257684
In[13]:= riemann[Sin, {0, Pi}, 16]
Out[13]= 1.99357
In[14]:= er[Sin, {0, Pi}, 16]
Out[14]= 0.00642966
```

Example 3. Similar to `riemann` and `er`, there are also defined functions to handle arbitrary partitions and arbitrary Riemann sums.[1]

```
In[15]:= ?arbitraryriemann
arbitraryriemann[f, P] calculates the Riemann sum of f for the partition
P by evaluating f at the right end point of each subinterval of P.
In[15]:= ?ear
ear[f, P] calculates the difference between the value of the defintite
    integral of f on the interval {Min[P], Max[P]} and
arbitraryriemann[f, P].
```

In addition there is a defined function `viewapprox` which allows us to see how good the Riemann sum approximation is:

```
In[15]:= ?viewapprox
viewapprox[f, P] shows the Riemann sum approximation to the definite
integral of f on the interval {Min[P], Max[P]} in terms of the given
partition P. The PlotLabel includes the function f, the number of
subintervals, the norm of the partition, and the absolute error of
the Riemann approximation. If P is a partition with some points
extremely close together - as may happen with a random partition -
the narrow rectangles on these short subintervals may be missed
in the output.
```

Note that in these functions, P can be taken to be any type of partition. For instance, we can repeat the calculations of In[11] simply by declaring P to be a regular partition of $[0, \pi]$ into 8 subintervals. And then we can see the approximation with `viewapprox`. See Figure 7.1.

---

[1]Although, as it stands, all the defined functions in this Chapter take the *right* end point of each subinterval.

7.1. PARTITIONS AND RIEMANN SUMS

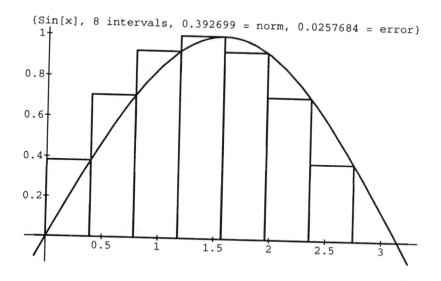

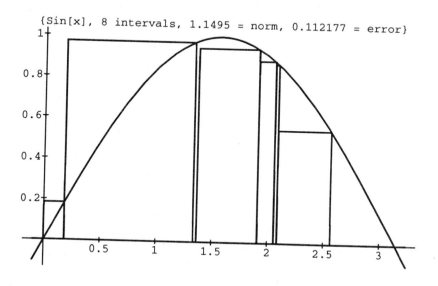

Figure 7.1: Approximating $\int_0^\pi \sin x\, dx$ with Riemann sums. Top: a regular partition of the interval; bottom: a random partition.

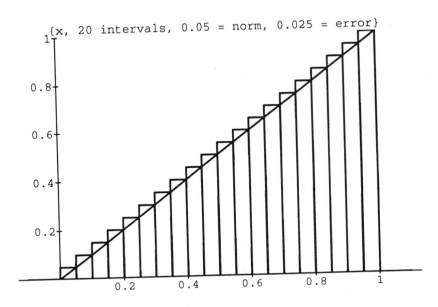

Figure 7.2: Approximating the area of a triangle with Riemann sums.

```
In[15]:= arbitraryriemann[Sin, regularpartition[{0, Pi}, 8]]
Out[15]= 1.97423
In[16]:= viewapprox[Sin, regularpartition[{0, Pi}, 8]];
```

For interest, we can compare with the results of using a random partition of the interval:

```
In[17]:= viewapprox[Sin, randompartition[{0, Pi}, 8]];
In[18]:= P
Out[18]= {0., 0.18544, 1.33494, 1.3671, 1.90834, 2.05725, 2.08826,
  2.56246, 3.14159}
```

See the bottom of Figure 7.1. Remember, you would obtain a totally different random partition to the one above.

Example 4. The calculation of areas is one of the main motivations behind Riemann sums. In this example we shall illustrate the process for the triangle bounded by the x–axis, $x = 0$, $x = 1$, and the line $y = x$. (The area is of course 1/2.) First we can tabulate a few Riemann sum approximations, and then we can view the process with **viewapprox**. See Figure 7.2.

```
In[19]:= f[x_] := x
```

7.1. PARTITIONS AND RIEMANN SUMS

```
In[20]:= Table[riemann[f, {0, 1}, n], {n, 20}]
Out[20]= {1., 0.75, 0.666667, 0.625, 0.6, 0.583333, 0.571429, 0.5625,
>       0.555556, 0.55, 0.545455, 0.541667, 0.538462, 0.535714, 0.533333,
>       0.53125, 0.529412, 0.527778, 0.526316, 0.525}
In[21]:= riemann[f, {0, 1}, 30]
Out[21]= 0.516667
In[22]:= viewapprox[f, regularpartition[{0, 1}, 20]];
```

This example was chosen because it is one of the few cases in which the limit of the Riemann sums can actually be evaluated directly. In this case, with $f(x) = x$, $a = 0$, and $b = 1$, the right side of Equation 7.1 reduces to

$$\sum_{i=1}^{n} \frac{i}{n^2},$$

which can be simplified by using the fact that

$$\sum_{i=1}^{n} i = \frac{n(n+1)}{2}.$$

Thus we have

$$\int_0^1 x\,dx = \lim_{n\to\infty} \frac{n(n+1)}{2n^2} = 1/2.$$

Example 5. In general the calculation of the limit of the Riemann sums is actually quite difficult. However, *Mathematica* offers a built in function, NIntegrate which gives a numerical approximation to any definite integral. We can consider it as a way to obtain the limit of the Riemann sums.

```
In[23]:= NIntegrate[f[x], {x, 0, 1}]
Out[23]= 0.5
```

MATHEMATICA FUNCTIONS INTRODUCED IN THIS SECTION
NIntegrate

Exercises:

1. Let $f(x) = x^3 - 3x + 2$ and consider the integral

$$T = \int_0^2 f(x)\,dx.$$

(a) Plot the graph of f on $[0, 2]$. By looking at the graph approximate the value of T.

(b) Find the exact value of T.

(c) Construct a table of Riemann sum approximations to T using regular partitions of $[0, 2]$ into $n = 2, 4, 6, 8, 10$, and 12 subintervals.

(d) View each of the approximations of part c) with `viewapprox`.

(e) Repeat part c) for *random* partitions of $[0, 2]$. Compare these approximations with those of part c).

(f) Compare the approximations visually for a regular partition and a random partition of $[0, 2]$ into 20 subintervals.

2. Consider the integral
$$\int_0^\pi \sin x \, dx.$$

(a) View the Riemann sum approximation for a regular partition of $[0, \pi]$ into 4 subintervals.

(b) Refine the partition of part a) by using `refine` to insert 5 additional points. View the Riemann sum approximation. How do the norm and the error compare with those from part a)?

(c) Choose any partition of $[0, 2]$ you wish and view the corresponding Riemann sum approximation. (If you are not sure how to do this, simply enter `viewapprox[Sin, { ...your points ...}]`.)

3. Consider the integral
$$\int_0^4 \arctan x \, dx.$$

(a) View the Riemann sum approximation for a regular partition of $[0, 4]$ into 6 subintervals. What is the error?

(b) Seven points are required to obtain a partition of six subintervals. Pick seven points to form a new partition of $[0, 4]$ to *decrease* the error in part a). Hint: consider the shape of the graph.

4. Use `NIntegrate` to evaluate the following definite integrals:

(a) $\int_0^3 e^{-x^2} dx$

(b) $\int_1^5 \frac{1}{x} dx$

(c) $\int_0^2 \cos(\cos x) dx$

(d) $\int_0^4 \sqrt{16 - x^2} dx$

5. For the integrals of the preceeding question, how large must n be in each case so that the Riemann sum approximation to the integral with a regular partition of n subintervals is within 0.1 of the answers from the preceeding question. (You can work visually with `viewapprox` or you can use the error term `er`.)

6. Let h be defined by
$$h(x) = \begin{cases} x^2 + 1, & \text{if } x \geq 0 \\ x, & \text{if } x < 0 \end{cases}$$
Note that h is not continuous at $x = 0$. Nevertheless,
$$\int_{-1}^{1} h(x)dx$$
still exists.

   (a) You can see this by viewing Riemann sum approximations with regular partitions of the interval $[-1, 1]$. Do this for $n = 10, 20, 30$, and 40. Describe what is happening at the discontinuity, and why the limit as $n \to \infty$ of the Riemann sums exists.

   (b) What is the exact value of the integral? Hint: add up the integrals of h over the two separate intervals $[-1, 0]$ and $[0, 1]$.

   (c) Use `NIntegrate` to evaluate the integral. Is this a very accurate answer?

7.2 The Fundamental Theorem of Calculus

The Fundamental Theorem of Calculus is the observation that if f is integrable on the interval $[a, b]$ then the function F defined by
$$F(x) = \int_a^x f(t)dt, \text{ for } a \leq x \leq b$$
is an antiderivative of f. That is, $F'(x) = f(x)$. Let us investigate this result graphically. We shall consider three examples. In the first example we shall use a function for which an antidervataive is well known; in the second example we shall use a funtion for which no elementary antiderivative is known; and for the final example we shall use a discontinuous function.

Example 1. Let f be defined by $f(x) = x^2$. This function has an obvious antiderivative of $x^3/3 + c$, for any c. Define F by
$$F(x) = \int_0^x f(t)dt.$$
We shall define F in *Mathematica* by making use of `NIntegrate`, as if we did not know the antiderivative of f. Then we shall plot F on the interval $[0, 2]$. See Figure 7.3.

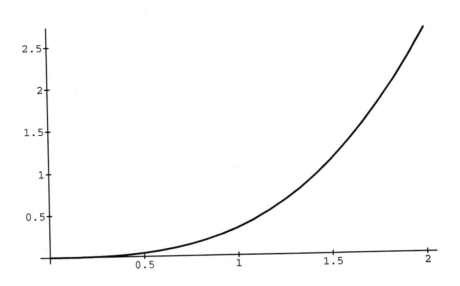

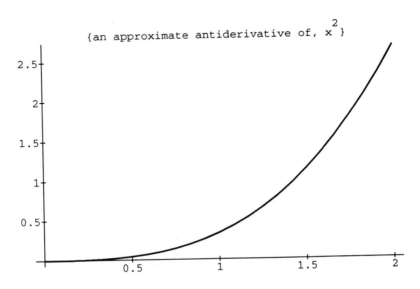

Figure 7.3: Top: the graph of $F(x) = \int_0^x t^2 dt$. Bottom: an antiderivative of $f(x) = x^2$.

7.2. THE FUNDAMENTAL THEOREM OF CALCULUS

```
In[1] := F[x_] := NIntegrate[t^2, {t, 0, x}]
In[2] := Plot[F[x], {x, 0, 2}, PlotRange -> All];
```

We can compare this graph with an antiderivative of f; suppose we use the function `viewantiderivative` which we defined in Chapter 6.

```
In[3] := << Chap6.m   (* Out[3] will be a
                                              usage statement *)
In[4] := f[x_] := x^2
In[5] := viewantiderivative[f, {0, 2}, 0];
In[6] := Show[%, Out[2]];
```

If you compare the two graphs, Figure 7.3, you can see that they are the same.

Example 2. We now repeat Example 1 for the function
$$f(x) = \cos(\cos x),$$
which you may recall cannot be integrated in any elementary way.

```
In[7]  := F[x_] := NIntegrate[Cos[Cos[t]], {t, 0, x}]
In[8]  := Plot[F[x], {x, 0, Pi}];
In[9]  := f[x_] := Cos[Cos[x]]
In[10] := viewantiderivative[f, {0, Pi}, 0];
In[11] := Show[%, Out[8]];
```

What we see, Figure 7.4, is that integrating f from 0 to x furnishes another way of constructing an antiderivative of f.

Example 3. As a final example, consider the function h defined by
$$h(x) = \begin{cases} x^2 + 1, & \text{if } x \geq 0 \\ x, & \text{if } x < 0 \end{cases}$$

This function is discontinuous at $x = 0$. Let us consider F to be defined by
$$F(x) = \int_{-2}^{x} h(t)dt, \text{ for } -2 \leq x \leq 2.$$

This way we shall be including the discontinuity of h in the domain of F. We proceed as before:

```
In[13] := h[x_] := If[x < 0, x, x^2 + 1]
In[14] := F[x_] := NIntegrate[h[t], {t, -2, x}]
In[15] := Plot[F[x], {x, -2, 2}];
Out[16] := (* ... a series of error messages regarding the
                   discontinuity of h[x] at x = 0 follow ... *)
In[16] := viewantiderivative[h, {-2, 2}, 0];
In[17] := Show[%, Out[15]];
```

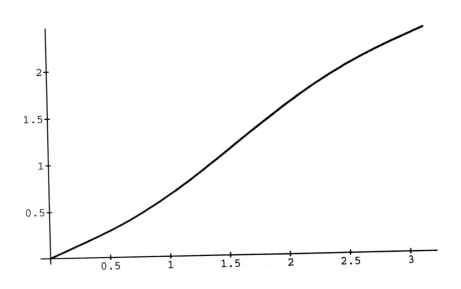

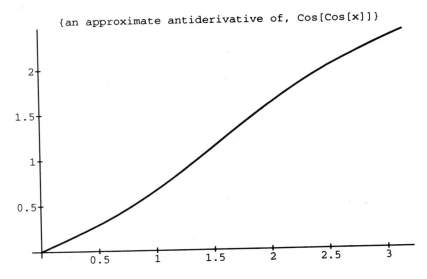

Figure 7.4: Top: the graph of $F(x) = \int_0^x \cos(\cos t)dt$. Bottom: an antiderivative of $f(x) = \cos(\cos x)$.

7.2. THE FUNDAMENTAL THEOREM OF CALCULUS

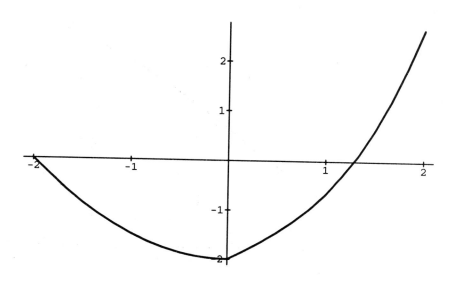

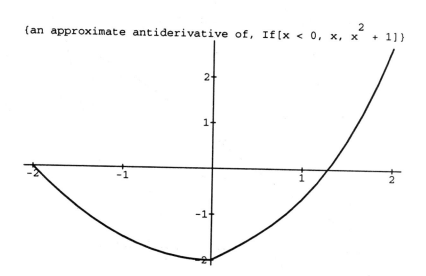

Figure 7.5: Top: the graph of $F(x) = \int_{-2}^{x} h(t)dt$. Bottom: an antiderivative of $h(x)$.

Again, we see that the two graphs are the same. (Figure 7.5.) What this, and the previous two examples show, is that there is a *geometrical* way to construct antiderivatives of any function by basically calculating the area under the given curve. (Of course, to do this we need to calculate the limit of Riemann sums, and this can cause *computational* problems — but the basic idea is *conceptually* simple. Certainly simpler than the method we developed in Chapter 6 to plot antiderivatives.)

Example 4. The Fundamental Theorem of Calculus is built in to *Mathematica*:

```
In[18]:= D[Integrate[f[t], {t, 0, x}], x]
Out[18]= f[x]
```

Also built in is the method of evaluating definite integrals if an antiderivative, F, of the integrand is known:

```
In[19]:= Integrate[D[F[x], x], {x, a, b}]
Out[19]= -F[a] + F[b]
```

Example 5. Evaluating a definite integral by finding an antiderivative often involves a change of variables (ie. a substitution) to simplify the integrand. In such a case, you should change the limits of integration as you change the variable. There is a defined function, definitetransform, in the package for Chapter 7 to do this:

```
In[20]:= << Chap7.m
 (* Out[18] will be a  usage statement *)
In[21]:= ?definitetransform
definitetransform[integrand, {x, a, b}, u, substitution] applies the
change of variables u = substitution (in terms of x) to the definite
integral of the given integrand (in terms of x) on the interval {a, b}
and outputs a list with three elements: the first element is the new
integrand (newintegrand) in terms of u (and possibly still x); the
second element is an equation giving u in terms of x; the third
element is a list of replacements of the new limits of integration
in terms of u.
```

We illustrate this function on the example

$$\int_0^4 x\sqrt{9+x^2}\,dx.$$

A standard substitution to use on this integral would be $u = 9 + x^2$.

```
In[21]:= definitetransform[x Sqrt[9 + x^2], {x, 0, 4}, u, 9 + x^2]
          Sqrt[u]             2
Out[21]= {-------, u == 9 + x , {9, 25}}
            2
```

7.2. THE FUNDAMENTAL THEOREM OF CALCULUS

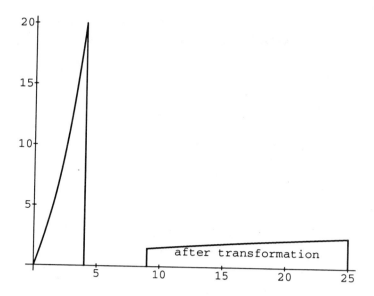

Figure 7.6: Transforming a definite integral. The area of the region is the same before and after transformation.

Note that the output gives the new integrand, in terms of u, and *the new limits of integration* in terms of u. With respect to u the problem is now

$$\int_9^{25} \frac{\sqrt{u}}{2} du.$$

It is of interest to plot the original integrand over the interval $[0, 4]$ and the new integrand over the interval $[9, 25]$. See Figure 7.6.

```
In[22]:= Plot[x Sqrt[9 + x^2], {x, 0, 4}, AspectRatio -> Automatic];
In[22]:= Plot[newintegrand, {u, 9, 25}, AspectRatio -> Automatic];
In[23]:= Show[Out[20], %];
```

As you can see, the area of the region that represents the value of the original definite integral appears equal to the area of the region that represents the new integral after the change of variables. (But this is not completely obvious, since the two regions have different shapes.)

FUNCTIONS CONTAINED IN THE PACKAGE Chap7.m		
regularpartition	randompartition	refine
riemann	arbitraryriemann	viewapprox
er	ear	norm
s	definitetransform	average

Exercises:

1. Find the derivative of the function
$$F(x) = \int_a^{x^3} \sin t\, dt.$$

2. Find the derivative of the function
$$F(x) = \int_{\sqrt{x^2+1}}^{b} e^t\, dt.$$

3. Generalize the results of Exercises 1 and 2 to the general case:
$$F(x) = \int_{h(x)}^{g(x)} f(t)\, dt.$$

    Try finding $F'(x)$ with *Mathematica* in two ways:

    (a) F'[x]

    (b) D[F[x], x]

    (c) Which, if any, of these methods results in an answer. Is it the correct answer?

4. Repeat Example 2 for the function
$$f(x) = (1 + x^2)^{1/3}$$
on the interval $[-3, 3]$.

5. Repeat Example 3 for the greatest integer function — Floor in *Mathematica* — on the interval $[0, 4]$. What do you notice about the antiderivative of Floor, in terms of continuity, even though Floor is itself not continuous?

6. Consider the two parabolas defined by f and g:
$$f(x) = 6x - x^2, \quad g(x) = x^2 - 2x.$$

    In this exercise we shall outline two ways to find the area of the region bounded by f and g.

7.2. THE FUNDAMENTAL THEOREM OF CALCULUS

(a) Plot f and g, find the intersection points, and set up the appropriate definite integral. Integrate with *Mathematica* to get the exact answer.

(b) Use `NIntegrate` on $|f(x) - g(x)|$, with the appropriate interval.

7. Repeat Exercise 6 for
$$f(x) = x^3 - 2x^2 + 1 \text{ and } g(x) = x.$$

(a) Which approach is more convenient in this case?

(b) Why is
$$\left| \int_{-0.801938}^{2.24698} (f(x) - g(x)) dx \right|$$
incorrect?

8. Find the area of the region bounded by the following curves:

(a) $f(x) = x^2 + 4x;\ g(x) = 0$
(b) $f(x) = x^2 - 4;\ g(x) = 8 - 2x^2$
(c) $f(x) = x^4 - 4x^2;\ g(x) = 4x^2$
(d) $f(x) = e^x;\ g(x) = e^{-x}$; on the interval $[0, e^3]$
(e) the two branches of $4(3x - y)^2 = 9x^3$ and $x = 4$
(f) $y = 9 - x^2;\ 4y = 5x^2;\ 64x = y^2$

9. Repeat Example 5 for the following definite integrals; use the indicated substitution:

(a) $\int_1^4 \frac{(1 + \sqrt{x})^4}{\sqrt{x}} dx;\ u = \sqrt{x}$

(b) $\int_0^{\sqrt{\pi}} x \sin(\frac{1}{2}x^2) dx;\ u = \frac{1}{2}x^2$

(c) $\int_0^{\pi/6} \sin 2x \cos^3 2x\, dx;\ u = \cos 2x$

10. The *average* value of f on the interval $[a, b]$ is defined as
$$\frac{\int_a^b f(x) dx}{b - a}.$$

There is a defined function in the package for Chapter 7 to compute averages; it is `average`.

(a) Query *Mathematica* about `average`.

(b) Use `average` to calculate the average of each of the following functions on the indicated interval:

i. $f(x) = mx + c$; on $[a, b]$
ii. $f(x) = x^2 + 3x - 3$; on $[0, 4]$
iii. $f(x) = \sin x$; on $[0, \pi]$
iv. $f(x) = x^3 + 4x - 5$; on $[0, 6]$

(c) For parts *ii*, *iii* and *iv* of part b), plot f and the constant function equal to the average of f, on the given interval. Use AspectRatio -> Automatic. The area of the rectangle with height equal to the average value of f should equal the area of the region between f and the x-axis.

11. The Mean Value Theorem for definite integrals states that if f is continuous on the interval $[a, b]$, then there is a number z such that

$$a < z < b \text{ and } \int_a^b f(x)dx = f(z)(b - a).$$

(This means that $f(z)$ is the average value of f.) Find z for the functions f in parts *ii*, *iii* and *iv* of the previous Exercise, part b).

12. Consider the function $f(x) = (1 + x^2)^{1/3}$ on the interval $[0, 10]$.

(a) What is the average value of f on $[0, 10]$?

(b) Find z in $[0, 10]$ such that $f(z)$ is the average value of f on $[0, 10]$.

(c) Finding z is equivalent to solving the equation

$$f(z) - \frac{\int_a^b f(x)dx}{b - a} = 0.$$

Find z by applying the *method of bisection* to this equation, as in Chapter 4, Section 2. (Since *Mathematica* cannot integrate f, use your numerical answer from part a) as the constant term in the above equation.)

Chapter 8

Numerical Integration

> *A victory is twice itself when the achiever brings home full numbers.* —W.Shakespeare, Much Ado about Nothing, 1598

8.1 The Trapezoid Rule and Simpson's Rule

As we have seen in the previous chapter, Riemann sums are not a practical way to calculate definite integrals: the Fundamental Theorem of Calculus provides a much more direct method of evaluating a definite integral, *if an antiderivative of the integrand is known.* However, it is a mathematical fact of life that not all elementary functions have elementary antiderivatives. Thus there is still a need for numerical methods to approximate some definite integrals. For instance, approximation is the only practical way to evaluate

$$\int_0^1 e^{-x^2} dx \text{ or } \int_2^5 (1+x^2)^{1/3} dx.$$

More surprisingly, perhaps, approximation may still be required *even if* the Fundamental Theorem can be used. After all, to know that

$$\int_0^1 \frac{4}{1+x^2} dx = \pi \text{ or that } \int_1^2 \frac{1}{x} dx = \log 2,$$

as we shall see later (or as we could find out directly by using *Mathematica*), still begs the question of the value of these integrals, since π and $\log 2$ are themselves irrational numbers and need somehow to be approximated. Indeed, one way of approximating such numbers is to approximate integrals! The trapezoid rule and Simpson's rule are two numerical methods for approximating definite integrals. In each case we take a regular partition of the interval $[a, b]$ into n subintervals:

In general, $x_i = a + \dfrac{i(b-a)}{n}$ and $x_i - x_{i-1} = \dfrac{b-a}{n}$. The trapezoid rule approximation of $\int_a^b f(x)dx$ is given by

(8.1) $$\int_a^b f(x)dx \simeq \dfrac{b-a}{2n}[f(x_0) + 2f(x_1) + 2f(x_2) + 2f(x_3) + \cdots + 2f(x_{n-1}) + f(x_n)]$$

If n is even, the Simpson's rule approximation of $\int_a^b f(x)dx$ is given by

(8.2) $$\int_a^b f(x)dx \simeq \dfrac{b-a}{3n}[f(x_0) + 4f(x_1) + 2f(x_2) + 4f(x_3) + \cdots + 2f(x_{n-2}) + 4f(x_{n-1}) + f(x_n)]$$

You will find that formulas for (8.1) and (8.2), as well as the formula for (7.1) are included in the package Chap8.m. They are **trapezoid**, **simpson**, and **riemann**.

```
In[1]:= << Chap8.m (* Out[1] is a long usage message *)
In[2]:= ?riemann
riemann[f, {a, b}, n] calculates the numerical value of the Riemann
sum obtained by dividing the interval {a, b} into n equally spaced
subintervals and evaluating f at the right hand end point of each
subinterval.
In[2]:= ?trapezoid
trapezoid[f, {a, b}, n] calculates the trapezoid rule approximation
of the definite integral of f on the interval {a, b} by dividing the
interval {a,b} into n equally spaced subintervals.
In[2]:= ?simpson
simpson[f, {a, b}, n] calculates the Simpson's rule approximation
of the definite integral of f on the interval {a, b} by dividing the
interval {a,b} into n equally spaced subintervals, where n is even.
```

Error expressions are included as well:

```
In[2]:= ?er
er[f, {a, b}, n] computes the absolute value of the difference
between the value of the definite integral of f on the interval
```

8.1. THE TRAPEZOID RULE AND SIMPSON'S RULE

{a, b} and the value of the Riemann sum approximation riemann[f, {a, b}, n].

```
In[2]:= ?et
```

et[f, {a, b}, n] computes the absolute value of the difference between the value of the definite integral of f on the interval {a, b} and the value of the trapezoid rule approximation trapdezoid[f, {a, b}, n].

```
In[2]:= ?es
```

es[f, {a, b}, n] computes the absolute value of the difference between the value of the definite integral of f on the interval {a, b} and the value of the Simpson's rule approximation simpson[f, {a, b}, n].

Example 1. In this example we shall compare the approximations offered by the trapezoid rule and Simpson's rule for the integral

$$\int_0^1 \frac{4}{1+x^2} dx = \pi.$$

```
In[2]:= f[x_] := 4/(1 + x^2)
In[3]:= trapezoid[f, {0, 1}, 4]
Out[3]= 3.13118
```

You can view graphs of the function f and your trapezoid rule approximation by using another defined function in **Chap8.m**, namely **viewtapprox**. See Figure 8.1, top part.

```
In[4]:= viewtapprox[f, {0, 1}, 4];
```

Note that the information across the top of the graph gives you the number of intervals used, the norm of the corresponding partition, and the absolute error. Similar functions are defined to view Riemann sum and Simpson's rule approximations, namely **viewrapprox** and **viewsapprox**. Now let's increase n and see what happens:

```
In[5]:= trapezoid[f, {0, 1}, 10]
Out[5]= 3.13993
```

Note that increasing n results in a better approximation. You should also view the approximation in this case to see how much closer to the curve the approximationg trapezoids actually are. See Figure 8.1, bottom part. If we now use Simpson's rule for the previous value of n, we obtain a much better approximation.

```
In[6]:= simpson[f, {0, 1}, 10]
Out[6]= 3.14159
```

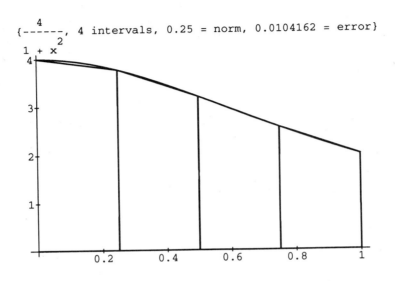

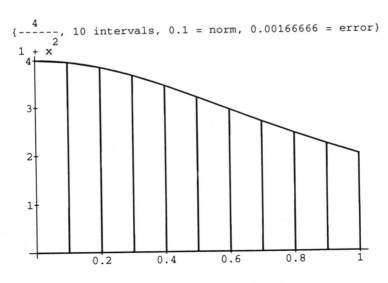

Figure 8.1: Trapezoid rule approximations to $\int_0^1 \frac{4}{1+x^2}dx$ with $n = 4$ (top), and with $n = 10$, (bottom).

8.1. THE TRAPEZOID RULE AND SIMPSON'S RULE

If you view the approximation via `viewsapprox[f, {0, 1}, 10]` you will notice that the approximating parabolas are practically indistinguishable from the graph of f itself. Indeed, this is practically true even for $n = 4$. See Figure 8.2. It is true in general that for a given value of n, Simpson's rule gives a better approximation than the Trapezoid rule. We can make this evident if we plot values of $\frac{1}{|error|}$ for different values of n, and compare the results for the different methods. Suppose, for interest, we generate such a plot for Riemann sum approximations first. See Figure 8.3, top left.

```
In[7]:= Table[1/er[f, {0, 1}, n], {n, 1, 10}]
Out[7]= {0.875969, 1.84641, 2.84213, 3.84001, 4.83871, 5.83784,
  6.83721, 7.83674, 8.83636, 9.83607}
In[8]:= ListPlot[%];
```

Then generate corresponding data for trapezoid rule approximations. See Figure 8.3, top right.

```
In[9]:= Table[1/et[f, {0, 1}, n], {n, 1, 10}]
Out[9]= {7.06251, 24.0427, 54.0081, 96.0045, 150.003, 216.002, 294.001,
>   384.001, 486.001, 600.001}
In[10]:= ListPlot[%];
```

Then for Simpson's rule approximations (remembering that n must be even). See Figure 8.3, bottom.

```
In[11]:= Table[{n, 1/es[f, {0, 1}, n]}, {n, 2, 10, 2}]
Out[11]={{2., 121.075}, {4., 41621.3}, {6., 1.1459310^6}, {8., 6.61677 10^6},
>   {10., 2.52203 10^7}}
In[12]:= ListPlot[%];
```

If you have kept all three graphs on your screen, or by looking at Figure 8.3, you can make a visual comparison of the data. For which method does the error term approach zero the quickest? for which method, the slowest? (RECALL: We are plotting $\frac{1}{|error|}$!) Are the plots in line with the theoretical "worst case" results

$$\text{er} \propto \frac{1}{n}, \quad \text{et} \propto \frac{1}{n^2}, \quad \text{es} \propto \frac{1}{n^4}?$$

Compare the data for Riemann sum and trapezoid rule approximations on the same graph:

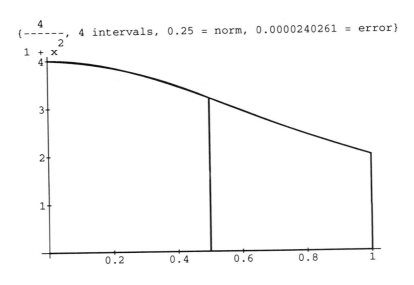

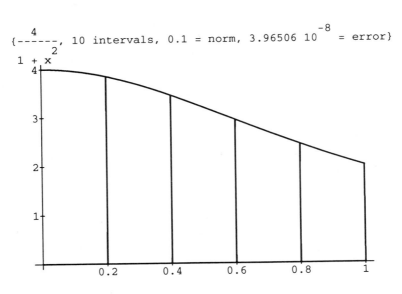

Figure 8.2: Simpson's rule approximations to $\int_0^1 \frac{4}{1+x^2}dx$ with $n = 4$ (top), and with $n = 10$, (bottom).

8.1. THE TRAPEZOID RULE AND SIMPSON'S RULE

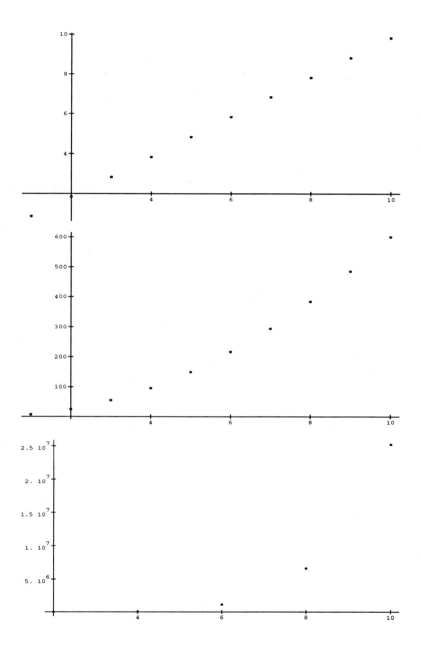

Figure 8.3: Plots of $\frac{1}{|error|}$ against the number of intervals, n, for Riemann sum, trapezoid rule and Simpson's rule approximations.

```
In[13]:= Show[Out[8], Out[10]];
```

Next compare the data for trapezoid and Simpson's rule approximations on the same graph:

```
In[14]:= Show[Out[10], Out[12]];
```

As you should be able to see upon comparing these three methods, Simpson's rule produces the best approximations, for given n.

Example 2. Using *Mathematica's* symbolic capabilities we can develop the trapezoid rule formula quite easily. We start with the simplest possible case — one interval $[a, b]$. We shall now find the equation $p(x) = s + tx$ of the linear function joining the points $(a, f(a))$ and $(b, f(b))$. Then we shall integrate $p(x)$ to obtain an approximation for $\int_a^b f(x)dx$. (Note that the Trapezoid Rule approximates the integral of f by first approximating f with p, and then integrating p — algebraically speaking. Geometrically, all that is necessary is to apply the formula for the area of a trapezoid — hence the name.)

```
In[15]:= Clear[f]
In[16]:= p[x_] := s + t x
In[17]:= Solve[{p[a] == f[a], p[b] == f[b]}, {s, t}]
                  b f[a]     a f[b]           f[a]        f[b]
Out[17]= {{s -> ------  -  ------,  t -> -(------) + ------}}
                 -a + b    -a + b          -a + b    -a + b
In[18]:= p[x] /. %[[1]]
           b f[a]    a f[b]            f[a]       f[b]
Out[18]= ------  -  ------  + x (-(------) + ------)
          -a + b    -a + b          -a + b    -a + b
In[19]:= Integrate[%, {x, a, b}]
                      2           2
           a b f[a]  b  f[a]   a  f[b]   a b f[b]
Out[19]= -(--------) + -------  + -------  -  --------  -
           -a + b     -a + b     -a + b     -a + b

             2   f[a]      f[b]        2   f[a]      f[b]
            a  (-(------) + ------)   b  (-(------) + ------)
                 -a + b    -a + b          -a + b    -a + b
    >     --------------------------- + ---------------------------
                        2                           2
In[20]:= Simplify[%]
           -((a - b) (f[a] + f[b]))
Out[20]= ------------------------
                     2
```

8.1. THE TRAPEZOID RULE AND SIMPSON'S RULE

Thus the trapezoid rule formula on one interval is

(8.3) $$\int_a^b f(x)dx \simeq \frac{(b-a)}{2}[f(a) + f(b)].$$

We can now move to the general case of n subintervals, each of length $(b-a)/n$, by applying (8.3) on each subinterval $[x_{i-1}, x_i]$, and adding up the results. This gives:

$$\begin{aligned}\int_a^b f(x)dx &\simeq \frac{(b-a)}{2n}[f(x_0) + f(x_1)] + \cdots + \frac{(b-a)}{2n}[f(x_{n-1}) + f(x_n)] \\ &= \frac{(b-a)}{2n}[f(x_0) + 2f(x_1) + 2f(x_2) + \cdots + 2f(x_{n-1}) + f(x_n)]\end{aligned}$$

Exercises:

1. Define a function f such that $f(x) = \frac{8x}{1+x^2}$. Consider the integral $X = \int_0^4 f(x)dx$.

   (a) Use **Integrate** and find the exact value of X.

   (b) Approximate the preceeding value using **% // N**.

   (c) Obtain an approximation of X using the trapezoid rule, with $n = 4$. Also view the approximation with **viewtapprox**.

   (d) Repeat part c) using Simpson's rule with $n = 4$.

   (e) Obtain trapezoid and Simpson's rule approximations for X with $n = 8$. View your results.

   (f) Do an error analysis, similar to that of Example 1, of Riemann sum, trapezoid rule, and Simpson's rule approximations of X.

2. Take the data for the error analysis of Example 1 and find the best polynomial fit to each of **Out[7]**, **Out[9]** and **Out[11]**. Use **Fit** to find polynomials that fit the data: by increasing the powers of x from 1 to 5 obtain five possible polynomials to fit the data in each case. Decide which polynomial fits the data best simply by looking at graphs including the polynomials and the points from **Out[7]**, **Out[9]**, and **Out[11]**. What are your conclusions?

3. This exercise outlines a development of the formula for Simpson's rule, similar to that of the trapezoid rule in Example 2.

   (a) Clear the definition of f and define a quadratic polynomial p such that
   $$p(x) = s + tx + ux^2.$$

(b) Solve for s, t, and u such that

$$p(a) = f(a), \quad p(\frac{a+b}{2}) = f(\frac{a+b}{2}), \quad p(b) = f(b).$$

Thus p will be the quadratic function that agrees with f on $[a, b]$ at a, b, and the mid-point $(a+b)/2$.

(c) Integrate p on the interval $[a, b]$ and simplify your answer. This will give you the simplest form of Simpson's rule, for $n = 2$.

(d) Now explain how to obtain (8.2) by applying the previous result to successive pairs of subintervals of $[a, b]$.

4. Even though *quadratic* approximations to f are used to obtain Simpson's rule, it actually gives precise answers for any *cubic* function f. (This is because the error estimate — see p 278, Edwards and Penney[3] – for Simpson's rule depends on the *fourth* derivative of f, and the fourth derivative of a cubic is zero.) You can observe this result symbolically with *Mathematica* as follows: define an arbitrary cubic polynomial $f(x) = s + tx + ux^2 + vx^3$, approximate $\int_a^b f(x)dx$ with the basic form of Simpson's rule obtained in Exercise 11, and compare this approximation with the exact value of $\int_a^b f(x)dx$.

8.2 Other Methods

If you compare the three simple methods of approximating integrals, namely Riemann sums, trapezoid rule and Simpson's rule, you will notice that the accuracy increases with number of points used to find the approximating functions to f. That is, the least accurate method, Riemann sum, uses one point on each subinterval to obtain a constant function approximation to f on each subinterval; the trapezoid rule uses two points on each subinterval to obtain the linear approximation to f on each subinterval; and Simpson's rule uses three points from two consecutive subintervals to obtain the quadratic approximation to f on each pair of subintervals. What would happen if you were to use *four* points from three consecutive subintervals to find a *cubic* approximation to f over three subintervals? Would the accuracy increase? We proceed to investigate this question using *Mathematica*.

Example 1. Define a cubic polynomial p; find its coefficients such that $p(x) = f(x)$ at

$$x = a, x = (2a+b)/3, x = (a+2b)/3, \text{and } x = b.$$

8.2. OTHER METHODS

$$a \quad\quad \frac{2a+b}{3} \quad\quad \frac{a+2b}{3} \quad\quad b$$

```
In[1]:= p[x_] := s + t x + u x^2 + v x^3
In[2]:= Solve[{p[a] == f[a], p[(a + 2b)/3] == f[(a + 2b)/3],
              p[(2a + b)/3] == f[(2a + b)/3], p[b] == f[b]},
              {s, t, u, v}]
Out[2]= ... pages of output scroll by ...

In[3]:= Simplify[%]
Out[3]= ... pages of output still scroll by ...

In[4]:= p[x] /. %[[1]]
Out[4]= ... the formula for p[x] scrolls by ...
In[5]:= Simplify[%]
Out[5]= ... its still a long formula for p[x] ...

In[6]:= Integrate[%, {x, a, b}]
Out[6]= ... about 7 pages of output scroll by ...
In[7]:= Simplify[%]
```

$$\text{Out[7]} = \frac{-((a-b)\,(f[a] + f[b] + 3\,f[\tfrac{2a+b}{3}] + 3\,f[\tfrac{a+2b}{3}]))}{8}$$

Let's call this expression the *cubic approximation* to $\int_a^b f(x)dx$. Here are some sample calculations to see how this new approximation compares with the trapezoid rule and Simpson's rule:

```
In[8]:= f[x_] := 4/(1 + x^2)
In[9]:= cubicapprox[f_, {a_, b_}] := (b - a) (f[a] + f[b] +
   3 f[(2 a + b)/3] + 3 f[(a + 2 b)/3])/8
In[10]:= cubicapprox[f, {0, 1}] // N
Out[10]= 3.13846
In[10]:= simpson[f,{0,1},2]  (* basic Simpson's rule with n = 2 *)
Out[10]= 3.13333
In[10]:= trapezoid[f,{0,1},1]  (* basic trapezoid rule with n = 1 *)
Out[10]= 3.
```

If you string together these cubic approximations on successive triples of subintervals of $[a, b]$ you will obtain a new numerical integration method known as *Simpson's Three Eighths Rule*. Some exercises below will allow you to investigate this rule a little more.

Example 2. *Mathematica's* built-in function for numerical integration is `NIntegrate`. By now you may be wondering what rule is behind `NIntegrate`? If you query *Mathematica* all you get is:

```
In[3]:= ?NIntegrate
NIntegrate[f, {x, xmin, xmax}] gives a numerical approximation to the
integral of f with respect to x over the interval xmin to xmax.
```

No particular rule is mentioned. In theory, Simpson's rule or the trapezoid rule could be used on a computer quite easily — n may need to be very large for some calculations, but after all, lots of calculations are what computers can do very quickly. However, there is a way to *reduce* the number of calculations required to achieve desired accuracy — it involves *adaptive* numerical integration, and `NIntegrate` uses it. We certainly won't go into the exact details used by `NIntegrate`, but we can illustrate the idea behind adaptive methods quite easily. Consider the graph of the function $y = \dfrac{8x}{1+x^2}$. Suppose we wish to approximate

$$\int_0^4 \frac{8x}{1+x^2}\,dx,$$

which for reference is equal to $4\log 17 \simeq 11.3329$. Let us use Riemann sums with $n = 6$:

```
In[11]:= f[x_] := (8 x)/(1 + x^2)
In[12]:= riemann[f, {0, 4}, 6]
Out[12]= 11.6311
```

If you look at the graph of f, Figure 8.4, you will notice that it fluctuates much more on the interval $[0, 2]$ than it does on $[2, 4]$. So instead of dividing each of these two intervals into 3 subintervals, let us divide $[0, 2]$ into 4 subintervals and $[2, 4]$ into 2 subintervals. This way we may get *more* accuracy than `Out[12]` *with the same amount of work*. Indeed:

```
In[13]:= riemann[f, {0, 2}, 4] + riemann[f, {2, 4}, 2]
Out[13]= 11.3285
```

Repeat this "adaptive" process for 10 subintervals. First use a Riemann sum with the 10 subintervals equally spaced:

8.2. OTHER METHODS

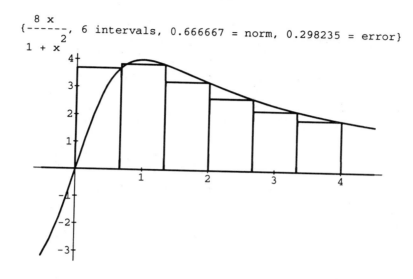

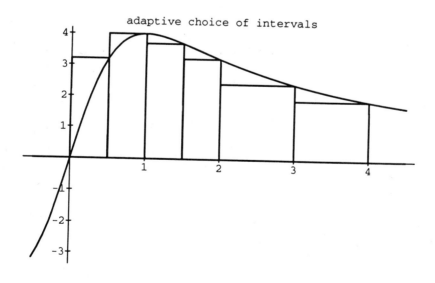

Figure 8.4: Adaptive Numerical Integration. After adapting the partition to the shape of the curve the error is only 0.00434659.

```
In[14] := riemann[f, {0, 4}, 10]
Out[14]= 11.5953
```

Now try adapting the number of subintervals to the shape of the graph:

```
In[15] := riemann[f, {0, 2}, 7] + riemann[f, {2, 4}, 3]
Out[15]= 11.3097
```

Again, the result is slightly more accurate than Out[14]. In this example Riemann sums were used because it is easier to compare visually a constant function with f, but for interest you should repeat the calculations of this example for trapezoid rule and Simpson's rule approximations.

FUNCTIONS CONTAINED IN THE PACKAGE Chap8.m		
riemann	trapezoid	simpson
er	et	es
s	l	q
viewrapprox	viewtapprox	viewsapprox

Exercises:

1. Define a function f such that

$$f(x) = \frac{-\sin x^2}{(1+\cos^2 x^2)^{3/2}}.$$

   Let us consider the integral $X = \int_0^2 f(x)\,dx$. (It has approximate value -0.56711.)

   (a) Approximate X by using the basic Simpson's rule, ie. with $n = 2$. View the accuracy of this approximation by using viewsapprox.

   (b) Approximate X by using the cubic approximation developed in Example 1. You will have to define the expression cubicapprox as in In[9] above.

2. Suppose the interval $[a, b]$ has been divided into n equally spaced subintervals, with n a multiple of 3. By summing up the cubic approximations for three subintervals of $[a, b]$ at a time, define a formula for Simpson's three eighths rule. Call it threeeighths.

3. Approximate X, as in Exercise 1 by using Simpson's rule with $n = 12$ and by using Simpson's three eighths rule with $n = 12$. Which is more accurate? Does this surprise you?

4. Similar to es, define an error term, ete, for Simpson's three eighths rule.

8.2. OTHER METHODS

(a) Make a table of $(n, \dfrac{1}{es[f, \{0, 2\}, n]})$ for n running from 2 to 36 in steps of 2, and `ListPlot` your data.

(b) For comparison, make a table of $(n, \dfrac{1}{ete[f, \{0, 2\}, n]})$ for n running from 3 to 36 in steps of 3, and `ListPlot` your data. Which seems to be the more accurate approximation method? Can you give any explanation for this?

5. Now we shall do an "adaptive" approximation of X.

   (a) Find the trapezoid rule approximation of X with $n = 6$; what is the absolute error? View your approximation wiht `viewtapprox`.

   (b) Instead of using 6 subintervals of $[0, 2]$ equally spaced, adapt the choice of intervals to the shape of the graph, to get a better approximation of X, still using the trapezoid rule.

   (c) Repeat parts a) and b), but now with $n = 12$ subintervals.

Chapter 9

Applications of the Integral

> *Curved areas are equal to each other if the ordinates are inversely proportional to the fluxions of the abscissas.* —I.Newton, Tractatus de Quadratura Curvarum 1704

The quote above is a statement in Newton's terms of his method of integration by substitution; the method of integration by parts had already been discovered by P. Fermat.

9.1 Integral Formulas

We now view the definite integral of a function as a limit of Riemann sums:

$$\int_a^b f(x)dx = \lim_{\text{Partition size} \to 0} \sum_{i=1}^n f(x_i^*)\Delta x_i.$$

This gives the area under the graph of f over the interval $[a, b]$. In later studies you will encounter new functions defined as integrals, often arising as the solutions to important differential equations. In many practical applications, the Riemann sums themselves arise as a result of making measurements with finite precision—every scientific instrument has limited precision and can only make estimates of average values over finite intervals of space and time. Typical integrals that arise in geometry and physics can be summarised as follows—and should be memorised:

$$\int_{x_1}^{x_2} A(x)\, dx = \text{Volume of solid with cross} - \text{section } A \text{ over } [x_1, x_2]$$

$$\int_{x_1}^{x_2} 2\pi x f(x)\, dx = \text{Volume of cylindrical solid with height f over } [x_1, x_2]$$

$$\int_{x_1}^{x_2} \pi r(x)^2\, dx = \text{Volume of solid of revolution with radius } r(x) \text{over } [x_1, x_2]$$

$$\int_{x_1}^{x_2} \sqrt{1 + (f'(x))^2}\, dx = \text{Arc length of curve } y = f(x) \text{ over } [x_1, x_2].$$

$$\int_{x_1}^{x_2} 2\pi f(x)\sqrt{1 + (f'(x))^2}\, dx = \text{Area from rotating curve } y = f(x) \text{ over } [x_1, x_2].$$

$$\int_{x_1}^{x_2} \text{Force}\, dx = \text{Work done moving particle over } [x_1, x_2]$$

$$\int_{t_1}^{t_2} \text{Velocity}\, dt = \text{Net movement over time interval } [t_1, t_2]$$

$$\int_{t_1}^{t_2} \text{Acceleration}\, dt = \text{Velocity change over time interval } [t_1, t_2]$$

$$\int_{t_1}^{t_2} \text{Flowrate}\, dt = \text{Quantity flowing over time interval } [t_1, t_2].$$

When trying to evaluate any integral it is worthwhile exploiting symmetries. As a trivial example of this, observe that the area enclosed by a circle is four times that under the curve defining one quadrant. Sometimes the volume you seek is more easily found by subtracting its complement from a larger simple region. The following sections illustrate some typical methods for computing volumes by integration. Often, more than one method may be used and a little thought and some preliminary sketches will help in choosing the best approach. Indeed, you can frequently make a rough guess at the volume to be found by making simplifying assumptions and this can save being embarassingly wrong through an algebraic slip! Of course, *all* methods eventually reduce to actually knowing the integrals of standard functions and standard procedures like integration by parts and through substitution.

9.2 Volume: Method of Cross Sections

This method arises by exploiting knowledge of the cross-sectional area $A(x)$ of a solid between the points $x = x_1$ and $x = x_2$:

$$\int_{x_1}^{x_2} A(x)\, dx = \text{Volume of solid with cross} - \text{section } A \text{ over } [x_1, x_2].$$

In *Mathematica* you will find some familiar shapes by inputting the package Shapes.m:

```
In[1]:=<< Graphics/Shapes.m
In[2]:=Show[Sphere[]];
```

which can help in visualizing the solids that they bound, see Figure 9.1. Observe how *Mathematica* renders these surfaces by means of finite numbers of congruent surface elements. This is often the same way that they are fabricated on a large scale in practice. It is instructive to sketch the family of the elements required for each of the surfaces in Figure 9.1. By imagining cuts along some joins between

9.2. VOLUME: METHOD OF CROSS SECTIONS

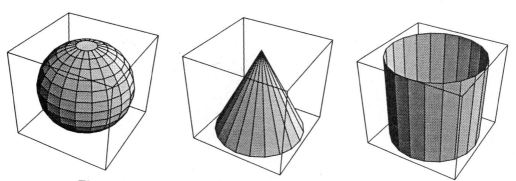

Figure 9.1: Approximations to the sphere, cone and cylinder

elements and laying them out flat in the plane but still connected, you can see how they could be made from one piece of material; sketch the limiting shape of this piece of material as you let the number of elements tend to infinity.

Exercises

1. In each case use *Mathematica* to plot the given cross-sectional area function A against x over suitable intervals. Estimate by eye from the plot the value of the integral, then compute it.

   (a) Sphere: $A(x) = \pi\sqrt{r^2 - x^2}$, $[-r, r]$.

   (b) Linear cone: $A(x) = \pi r^2 x$, $[0, h]$.

   (c) Cubic cone: $A(x) = \pi r^2 x^3$, $[0, h]$.

2. By considering the way *Mathematica* produces the graphic approximation to a linear cone in Figure 9.1 as a pyramid with a finite number of sides, find a formula for its approximate cross-sectional area in terms of n the number of flat triangles used to approximate the cone. What happens as n tends to infinity? Can you find the corresponding formula and limit for the volume of pyramids with n sides? You might try first the cylinder. Generate some other surfaces using flat triangular elements and investigate the limiting formulas for area and volume.

3. The work done in pumping fluid of density ρ from a level $y = 0$, into a tank of cross-sectional area $A(y)$ at height y with $y_1 \leq y \leq y_2$, is given by:

$$\int_{y_1}^{y_2} \rho y A(y) \, dy.$$

Compute this work for rectangular, spherical, conical and cylindrical tanks of similar volumes. First estimate these answers by suitable simplifying assumptions.

4. The work done in pumping fluid of density ρ to a height $h > b$, from a tank of cross-sectional area $A(y)$ at height y with $a \leq y \leq b$, is given by:

$$\int_a^b \rho(h-y)A(y)\,dy.$$

Compute this work for spherical, conical and cylindrical tanks. First estimate these answers by suitable simplifying assumptions.

5. The total force on a submerged vertical plate of width $w(y)$ at height y in a tank of fluid of density ρ is

$$\int_a^b \rho(c-y)w(y)\,dy,$$

where the plate lies in the interval $a \leq y \leq b$ and the top of the fluid is at level $y = c > b$. Compute this force for small circular inspection plates in the sides of rectangular, spherical, conical and cylindrical tanks.

9.3 Solid of Revolution

This is a special case of the method of cross-sections. If a solid is constructed by revolving around the x-axis a region between the two curves $y = f(x)$ and $y = g(x)$, over interval $[x_1, x_2]$, then the cross-sectional area function is given by $A(x) = \pi(f(x) - g(x))^2$. There is a possible generalization to revolution about another axis; this simply involves some algebra to relocate the origins for the bounding curves.

Exercises

1. In each case use *Mathematica* to plot the given functions f, g against x over the interval. If no interval is given, then solve the equation $f(x) = g(x)$ to find the interval for integration. Estimate by eye from the plot the volume obtained by revolving the region bounded by f, g around the x-axis (or another if given), then compute it. Use Plot3D to show the object being studied.

   (a) Punctured sphere: $f(x) = \sqrt{1-x^2}$, $g(x) = \frac{1}{2}$, $[-1, 1]$.

   (b) Parabolic bounds: $f(x) = \sqrt{x}$, $g(x) = \sqrt{2(x-3)}$.

   (c) $f(x) = \sqrt{x}$, $g(x) = x^3$; about the x-axis, about the y-axis and about $x = -1$.

   (d) $f(x) = x^2$, $g(x) = 4x$, about $x = 5$.

9.4. VOLUME: METHOD OF CYLINDRICAL SHELLS

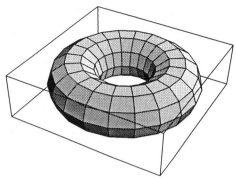

Figure 9.2: Torus, formed by joining the ends of a cylinder or tube

2. Find the volume of a spherical cap of height h in a sphere of radius r with $h < r$.

3. Estimate the volume of cream needed for a typical spherical cream donut.

4. Find the volume of a torus formed by revolving a circle around the y-axis.

5. Estimate the volume of chocolate needed to cover the top half of a typical donut of torus type to a depth of 0.5 mm, see Figure 9.2.

6. Find the volume of the solid formed by revolving the ellipse
$$\frac{x^2}{a^2} + \frac{y^2}{b^2} = 1$$
round its major axis. Repeat the problem for revolution about the minor axis.

9.4 Volume: Method of Cylindrical Shells

This method arises by exploiting the fact that a volume to be computed has cylindrical symmetry but a possibly varying height function f. Thus, $2\pi x$ is the perimeter of a circle of radius x, and when multiplied by $f(x)$ it gives the area of a cylindrical shell of that height. Integrating this over the domain of f gives the required volume:

$$\int_{x_1}^{x_2} 2\pi x f(x)\, dx = \text{Volume of cylindrical solid with height } f \text{ over } [x_1, x_2]$$

Exercises

In each case use *Mathematica* to plot the given height function f against x over the interval. Estimate by eye from the plot the value of the integral, then compute it. Use Plot3D to show the object being studied.

1. $f(x) = 3 - x$, $[1, 2]$
2. $f(x) = 9 - x^2$, $[1, 2]$
3. $f(x) = \sqrt{1 - x^2}$, $[0, 1]$
4. $f(x) = \sqrt{1 - x^2}$, $[\frac{1}{2}, 1]$

9.5 Arc Length

$$\int_{x_1}^{x_2} \sqrt{1 + (f'(x))^2} \, dx = \text{Arc length of curve } y = f(x) \text{ over } [x_1, x_2].$$

If the curve is given in the form $x = g(y)$, over $[y_1, y_2]$, then the arc length is:

$$\int_{y_1}^{y_2} \sqrt{1 + (g'(y))^2} \, dy = \text{Arc length of curve } x = g(y) \text{ over } [y_1, y_2].$$

The integrals arising from the formula for arc length can be very difficult to evaluate analytically but can always be approximated numerically. Firstly, however, look at a Möbius strip, Figure 9.3 by inputting

In[3]:=Show[MoebiusStrip[]];

Given that it is made by cutting and rejoining with a twist a cylinder of height h and radius r, what is the arc length of its single edge? What is its surface area? See Figure 9.3 for finite rectilinear approximations.

Exercises

1. In each case use *Mathematica* to draw the curve over the interval. Estimate by eye from the plot the length of the curve, then compute it.

    (a) $y = x^{\frac{3}{2}}$, $[0, 5]$.
    (b) $x = \frac{y^3}{6} + \frac{1}{2y}$, $[1, 2]$.
    (c) $y = 2x^{\frac{2}{3}}$, $[1, 8]$.

2. Use ParametricPlot to draw the closed curves $x^n + y^n = 1$, for even $n = 2, 4, 6, \ldots$. Write down integrals which express the arc lengths of these curves and evaluate them numerically. Use ListPlot to plot the arc lengths as a function of n. First, of course, you must find the range of values taken by one of the variables, x say; your knowledge of the case $n = 2$ should help.

9.5. ARC LENGTH

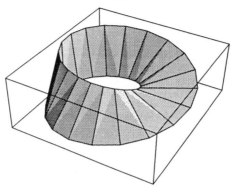

Figure 9.3: Möbius strip, made from a flat strip joined at the ends after one twist; it has one edge and one surface

3. Write down an integral to express the arc length of the curve $y = \sin 2\pi n x$, for x in the interval $[0, 1]$. For $n = 0, 1, 2, \ldots$, compute the numerical value of this integral and use ListPlot to show how it depends on the frequency n of the sine wave.

4. Show that an ellipse can be presented parametrically in the form
$$x = a\cos\theta, \quad y = b\sin\theta$$
and deduce that the element of arc length can be expressed by
$$ds = (a^2 \cos^2\theta + b^2 \sin^2\theta)^{\frac{1}{2}} d\theta.$$
Hence deduce the perimeter of the ellipse.

5. In polar coordinates the element of arc length takes the form
$$ds^2 = dr^2 + r^2 d\theta^2.$$
Use this to find the lengths of the following classical curves. We indicate how to plot some of the curves by giving the *Mathematica* input. The results are shown in Figure 9.4.

(a) The spiral of Archimedes, $r = a\theta$, and its inverse, the reciprocal spiral $r\theta = a$. Show that the spiral length measured from its pole is the same as that of a parabola $y^2 = 2ax$ measured from its vertex.

```
In[4]:=ParametricPlot[{(u)Cos[u],(u)Sin[u]},
  {u,0,6Pi}, PlotLabel->"Spiral of Archimedes"];
```

```
In[5]:=ParametricPlot[{(1+ Cos[u])Cos[u], (1+ Cos[u])Sin[u]},
        {u,0,2Pi}, PlotLabel->"Cardioid"];

In[6]:=ParametricPlot[{(1+2Cos[u])Cos[u], (1+2Cos[u])Sin[u]},
        {u,0,2Pi}, PlotLabel->"Limacon of Pascal"];
```

(b) The cardioid, $r = a(1 + \cos\theta)$.

(c) The limacon of Pascal, $r = 1 + 2\cos\theta$.

(d) The lituus, $r = a\theta^{-\frac{1}{2}}$.

9.6 Surface Area of Revolution

It is quite common to have to find the area of a surface generated by the rotation of a given curve about some axis. A small element of arc length Δs rotated about an axis produces an annulus of approximate area $2\pi r \Delta s$, if the axis is distance r from the element of arc. Note that, conventionally, we presume unless told otherwise that we want only the area of *one* side of the surface. For a curve $y = f(x)$ rotated about the x-axis we have:

$$\int_{x_1}^{x_2} 2\pi f(x)\sqrt{1+(f'(x))^2}\,dx = \text{Area from } x\text{-rotating curve } y = f(x) \text{ over } [x_1, x_2].$$

If rotated about the y-axis then the formula is:

$$\int_{x_1}^{x_2} 2\pi x\sqrt{1+(f'(x))^2}\,dx = \text{Area from } y\text{-rotating curve } y = f(x) \text{ over } [x_1, x_2].$$

The two formulae are easily remembered in the forms:

$$\int_{x_1}^{x_2} 2\pi f(x)\,ds, \quad \int_{x_1}^{x_2} 2\pi x\,ds, \quad \text{respectively.}$$

Exercises

1. Write down the corresponding formulae for surfaces generated by curves presented in the form $x = g(y)$.

2. In each case use *Mathematica* to draw the curve over the interval. Estimate by eye from the plot the length of the curve and the area generated by revolution about the given axis over the specified interval, then compute it. Use ParametricPlot3D to illustrate the surfaces produced.

    (a) $y = \sqrt{x}$, $[0,1]$ about the x-axis and about the y-axis.

    (b) $y = (3x)^{\frac{1}{3}}$, $[0,9]$ about the y-axis.

9.6. SURFACE AREA OF REVOLUTION 217

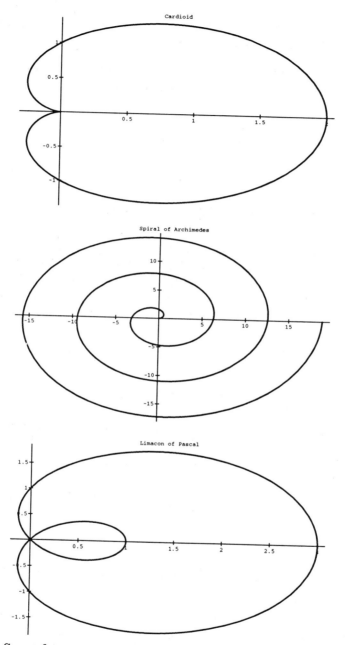

Figure 9.4: Some famous curves: the cardioid, the spiral of Archimedes and the limacon

3. What happens if you rotate a segment of a curve with a twist—consider the Möbius strip and cylinder.

4. Use *Mathematica* to find numerically the surface areas of revolution generated by rotating the following curves about the x-axis and about the y-axis; plot the curves and surfaces.

    (a) $y = \sin 2\pi nx$, $[0,1]$, for $n = 1, 2, 3$.
    (b) $y = (1 - x^{\frac{2}{3}})^{\frac{3}{2}}$, $[-1,1]$.
    (c) $(x-4)^n + (y-4)^n = 1$, for $n = 2, 4, 6$.
    (d) $(x-4)^2 + 2(y-4)^2$.

9.7 Separable Differential Equations

You can think of a differential equation

$$\frac{dy}{dx} = F(x,y)$$

as a way to describe a family of curves in the xy-plane. It is a remarkable result of calculus that if F is nicely behaved then there is *one and only one* curve going through any given point (x_0, y_0) in the plane. Thus, if we are given a family of starting points for our differential equation, then we can deduce the corresponding family of curves—provided that we can solve the equation.

There is an important special case when the prospect for analytic solution becomes brighter: when it is possible to separate $F(x,y)$ into a product of two functions, one in x only and the other in y only; in this case we say that the equation is *separable*. Beware that sometimes you need to do a little algebra to establish such separability and a change of variable may help. It is always good practice to look at the tangent direction of the curve at its initial point, to see where it is going and at what rate. Indeed, the repeated use of this to develop an approximate solution curve is at the heart of the proof of its existence, and vital to the numerical integration algorithms.

Exercises

1. Plot the family of curves $y = x^2 + c$ for $0 \le x \le 1$ and $c = 0, 0.1, 0.2, \ldots, 1.0$. Construct a separable first order differential equation for which this family consists of solutions with $x_0 = c$. Repeat this for some other curves that you have met—what about a family of concentric circles?

2. Find and plot the families of solution curves for the following separable equations:

(a) $\frac{dy}{dx} = xy$, $y(0) = 0, 0.1, 0.2, \ldots, 1.0$.

(b) $\frac{dy}{dx} = xy^2$, $y(0) = 0, 0.1, 0.2, \ldots, 1.0$.

(c) $\frac{dy}{dx} = x^2 y$, $y(0) = 0, 0.1, 0.2, \ldots, 1.0$.

(d) $\frac{dy}{dx} = (xy)^{\frac{1}{2}}$, $y(0) = 0, 0.1, 0.2, \ldots, 1.0$.

(e) $\frac{dy}{dx} = (xy)^{\frac{1}{3}}$, $y(0) = 0, 0.1, 0.2, \ldots, 1.0$.

(f) $\frac{dy}{dx} = (xy)^{\frac{1}{4}}$, $y(0) = 0, 0.1, 0.2, \ldots, 1.0$.

3. A particle moving in the xy-plane can have its trajectory represented parametrically by a curve, in terms of the parameter time, t. Thus $y = r \sin t$, $x = r \cos t$, represents motion in a circle of radius r. Construct the corresponding differential equation to generate a family of such trajectories indexed by r. Repeat this for other closed curves, such as ellipses and the 'circles with corners' : $x^n + y^n = r^n$, for even n and $r = 0, 0.1, 0.2, \ldots, 1.0$.

Chapter 10

Exponential and Logarithmic Functions

Eadem mutata resurgo. —J.Bernouilli, on his tombstone with the graph of the logarithmic spiral $\log r = a\theta$, 1705

10.1 The Natural Logarithm as a Definite Integral

You have already met the exponential and natural logarithm functions informally; here we take a brief, somewhat more sophisticated approach to them as important consequences of the way calculus is constructed. In fact, it is fair to claim that the exponential function is the single most important function in all mathematics; the natural logarithm is its inverse function. We define the latter as an apparently simple integral, then switch on the theorems. The definition of functions in terms of integrals of simpler functions is actually rather common in mathematics and in its applications to science and engineering.

For $x > 0$, the natural logarithm is the function defined by

$$\log x = \int_1^x \frac{1}{t}\, dt.$$

We can plot this integral as a function of its upper limit as follows:

```
In[1]:=Integrate[1/t,{t,1,x}]
Out[1]:=Log[x]
In[2]:= Plot[Integrate[1/t,{t,1,x}],{x,0.1,10},
                AxesLabel->{"x","Log[x] "}];
```

221

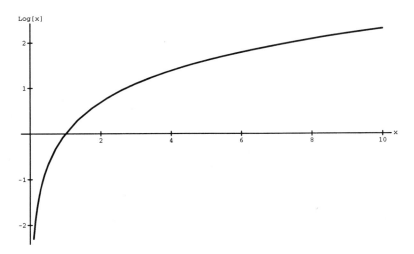

Figure 10.1: Plot of the integral $\int_1^x 1/t\,dt$ as a function of its upper limit

The result is shown in Figure 10.1. Why this integral should be singled out is not so obvious, but note that integrals of other integer powers of a variable are expressible in terms of existing functions, and we have to stay away from zero if we use reciprocals. Immediately we deduce that $\log 1 = 0$, by definition of the integral, and that log is continuous. Moreover, the Fundamental Theorem of Calculus gives us its derivative:

$$\frac{d}{dx}\log x = \frac{1}{x}.$$

Since this derivative is positive for $x > 0$, we conclude that log is a strictly increasing function, hence has an inverse. We shall call this inverse *exp* and investigate some properties in the next section. We observe two things from the graph of log : its inverse *exp appears* to be defined over the whole real line and of course is never negative; it is shown in Figure 10.2. Also, if we use the modulus function then $\log|x|$ is well-defined for all $x \neq 0$.

In[3]:=Plot[E^x ,{x,-2 ,3 },AxesLabel->{"x","E[x]"}];

By definition then:

$$exp(x) = y \quad \text{if and only if } \log y = x,$$

and so it turns out that

$$\log(exp(x)) = x \quad \text{for all } x;$$
$$exp(\log y) = y \quad \text{for all } y > 0.$$

Exercises

10.2. THE NATURAL EXPONENTIAL FUNCTION

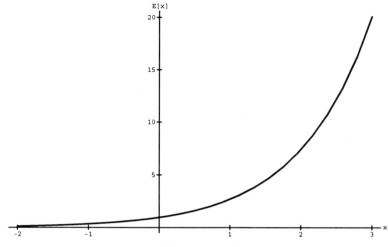

Figure 10.2: Plot of *exp x*

1. The following properties of the log function are the ones you will use most often:

    (a) $\log xy + \log x + \log y$.
    (b) $\log \frac{1}{x} = -\log x$.
    (c) $\log \frac{x}{y} = \log x - \log y$.
    (d) $\log x^r = r \log x$.
    (e) $\frac{d}{dx} \log |u(x)| = \frac{1}{u(x)} \frac{d}{dx} u(x)$, for any differentiable function u that is never zero. This is an application of the chain rule; try some familiar functions in place of u.

2. Plot the ratio $\frac{\log x}{x^n}$ against x for a range of values $n = 1, 2, \ldots$.

3. Use *Mathematica* to investigate the fact that for all $n = 1, 2, \ldots$

$$\lim_{x \to \infty} \frac{\log x}{x^n} = 0.$$

4. Investigate the power series expansion of $\log(1 + x)$ for x near zero using Series:

```
In[1] := Series[Log[1+x], {x,0,10}].
```

10.2 The Natural Exponential Function

Since $\log 1 = 0$ and log is increasing, somewhere we must have a point $x = e$, say, at which $\log x = 1$. This turns out to be a very interesting number and in mathematics the symbol e always means this number; in value it is about 2.71828 if you ask *Mathematica* for the numerical solution to the above equation. The algebraic solution in *Mathematica* is E, in keeping with its use of capitals for internal definition of functions—for this E stands for the exponentiation function, as we shall see.

From the properties of log we deduce that

$$\log e^x = x \log e = x.$$

But we already have that

$$\log(exp(x)) = x$$

and inverse functions are unique so

$$exp(x) = e^x, \quad \text{for all } x.$$

This is a non-trivial result because for the first time we have the possibility to use *irrational* powers also. The formal definitions are summarised by:

$$a^x = e^{x \log a} \quad \text{for all } x, \text{ and all } a > 0$$
$$x^r = e^{r \log x} \quad \text{for all } x, r.$$

Exercises

1. The following properties of the exponential function are the ones you will use most often:

    (a) $e^x e^y = e^{x+y}$

    (b) $e^{-x} = \frac{1}{e^x}$, $(e^x)^r = e^{rx}$

    (c) $\frac{d}{dx} e^x = e^x$

    (d) $\frac{d}{dx} e^{u(x)} = e^{u(x)} \frac{du}{dx}$.

2. Investigate the power series expansion of e^x for x near zero using **Series**:

    In[1] := Series[Exp[x], {x,0,10}].

    Guess the first two terms of the power series of the inverse function of e^x.

3. Guess the values of $\sqrt{2}^{\sqrt{2}}$, π^π, e^e, and then compute them.

4. Plot the ratio $\frac{x^n}{e^x}$ against x for a range of values $n = 1, 2, \ldots$.

5. Use *Mathematica* to investigate the fact that for all $n = 1, 2, \ldots$

$$\lim_{x \to \infty} \frac{x^n}{e^x} = 0.$$

10.2. THE NATURAL EXPONENTIAL FUNCTION

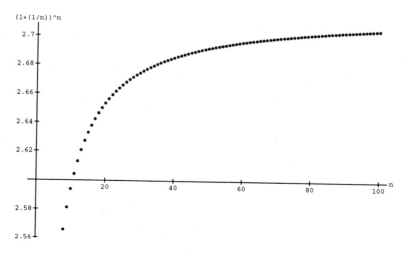

Figure 10.3: Plot of $(1 + (1/n))^n$

6. Plot the values of the function $(1 + \frac{1}{n})^n$ for $n = 1, 2, \ldots$. The graph is shown in Figure 10.3. Investigate the facts that

$$\lim_{n \to \infty} (1 + \frac{1}{n})^n = e$$
$$\lim_{n \to \infty} (1 + \frac{x}{n})^n = e^x \text{ for all } x.$$

We can illustrate the procedure for the first of these:

```
In[4]:=ListPlot[Table[(1+(1/n))^n ,{n,1,100}],
              AxesLabel->{"n","(1+(1/n))^n"}];
```

Now, if we ask *Mathematica* the numerical value of e, it tells us 2.71828 to 6 digits. We can draw this level in our previous graph as follows. Firstly,

```
In[5]:=Plot[2.71828,{x,0,100}];
```

draws the constant function 2.71828, (which we omit) then

```
In[5]:=Show[%,%%];
```

puts both together, as in Figure 10.4.

7. Use the Binomial Theorem (or your knowledge of *all* the derivatives) to obtain a power series expansion for e^x :

$$1 + \frac{x}{1!} + \frac{x^2}{2!} + \frac{x^3}{3!} + \cdots$$

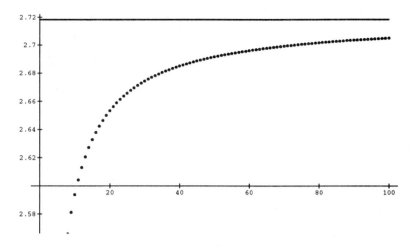

Figure 10.4: $(1 + (1/n))^n$ tends (slowly!) to e

and investigate how quickly it approaches e^x for different x. Do this by using the ListPlot[Table[ ]] function in *Mathematica* to plot the value of this series up to order n against n for each value of x. If the series is really correct, then it should equal its own derivative and its own indefinite integral, shouldn't it? Can you find any other series with that property?

10.3 Applications

You will find numerous examples in the course text and in your other studies to illustrate the ubiquity of the exponential function. Typically in these examples, a function of time, N say, increases or decreases in proportion to its current value:

$$\frac{d}{dt}N = kN, \quad \text{with initial value } N(0) = N_0.$$

This is a first order differential equation and we can easily solve it to find

$$N(t) = N_0 e^{kt},$$

representing exponential growth or decay according as k is positive or negative. A minor generalization is to the case that there is an additional term on the right hand side of the differential equation, which gives rise to the general first order linear differential equation:

$$\frac{d}{dt}x = ax + b$$

10.3. APPLICATIONS

for some functions a, b of t only. When a and b are *constants* then the solution to this is

$$x(t) = (x_0 + \frac{b}{a})e^{at} - \frac{b}{a},$$

with $x(0) = x_0$.

Exercises

1. Find examples illustrating the use of the exponential function to model growth and decay. In each case use *Mathematica* to plot families of solution curves to show the effects of possible errors in the given initial data or constants for the model.

2. Probability theorists often model random processes by the *Poisson Distribution*:

$$p(n) = \frac{m^n}{n!} e^{-m}$$

wherein $p(n)$ is the probability that an event will happen precisely n times in a given interval (of time or space). For example, if you use a random number generator to provide coordinate pairs of points on a square grid, the probability that individual subsquares contain specified numbers of points is well modelled by such a distribution; the expected or mean number per subsquare is just the ratio of the total number of points to the number of subsquares. Plot the Poisson distribution using ListPlot, for a range of values $m = 1, 2, 3, 4$. We give the plot for $m = 2$:

```
In[6]:= Table[E^-2 2^n/n! ,{n,0,10 }]
In[7]:= ListPlot[%,AxesLabel->{"n","Poisson Distribution"}];
```

ListPlot is designed to plot over values $n = 1, 2, \ldots$, so the values would be displaced if you used ListPlot[Table] for $n = 0, 1, \ldots$.

Observe how the probability of no event happening ($p(0)$) changes with the Poisson parameter m. There is a maximum value if $m > 1$ and the distribution becomes increasingly symmetric as m increases above 1. Probabilistically, the mean number of times an event happens is the sum of the possible numbers multiplied by their respective probabilities:

$$\text{Mean number of events } = 0p(0) + 1p(1) + 2p(2) + 3p(3) + \cdots.$$

Investigate this sum for some small integer values of m. The variance of the number of events is the difference between the mean square and the square of the mean:

$$\text{Variance of } n = \text{Mean of } n^2 - (\text{Mean of } n)^2.$$

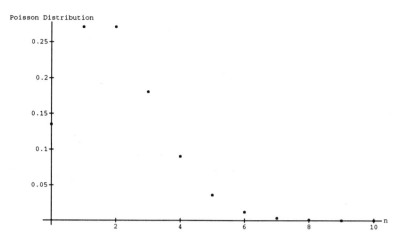

Figure 10.5: Plot of the Poisson Distribution $e^{-2}\frac{2^n}{n!}$

Algebraically, the variance is given by:

$$0^2p(0) + 1^2p(1) + 2^2p(2) + 3^2p(3) + \cdots - (0p(0) + 1p(1) + 2p(2) + 3p(3) + \cdots)^2.$$

Investigate this for a range of values $m = 1, 2, 3, 4$. Can you show algebraically what the result of the infinite sums are? It will help to know the power series for the exponential function:

$$e^m = \frac{m^0}{0!} + \frac{m^1}{1!} + \frac{m^2}{2!} + \frac{m^3}{3!} + \cdots.$$

3. Use the log function on the equation

$$y = kx^m, \quad \text{for positive } y, k, x$$

to express the relationship between y and $\log x$.

Chapter 11

Trigonometric and Hyperbolic Functions

> *The ratio of the side opposite angle A to the hypotenuse of a triangle must be the same in any right-angled triangle containing angle A.* —Hipparchus, inventing sin A and making the first sine tables, 180-125 BC

11.1 Motivation

Algebraic problems gave rise to the construction of new numbers: polynomial problems led to the invention of irrationals like $\sqrt{2}$ and complex numbers. Calculus problems give rise to new functions, and new interpretations of very old functions like sine . We noted how intricately were the properties of the exponential function tied up with the properties of calculus itself: it (or any constant multiple of course) is the only function (except the constant 0) which is its own derivative. So, e^x is essentially the only interesting invariant of calculus:

$$\frac{d}{dx}f = f \quad \text{if and only if} \quad f(x) = ae^x, \text{ for some constant } a.$$

We have also solved the problem of what function is the negative of its own derivative, e^{-x}. It is now of interest to know the functions which satisfy the next level of invariance:

$$\frac{d^2}{dx^2}f = \pm f.$$

The solutions to this problem consist of some trigonometric and hyperbolic functions, which also have power series expressions that are valid for all real numbers. We illustrate some of their properties in this Chapter. Like the exponential function, the trigonometric and hyperbolic functions are *transcendental* in the sense that they do not arise from purely algebraic operations—they involve limits of infinite series.

These are also well-defined for complex numbers but you will not meet that aspect until a later course. First we look at some limits involving trigonometric functions.

11.2 The Area of a Unit Circle

We know that the area of a unit circle is π. Now we can use trigonometry to show how it can be estimated from upper and lower bounds which are being forced together. Consider the exscribed square and the inscribed square of the unit circle. These give upper and lower bounds for the area of the circle. Now replace the squares successively by pentagons, hexagons, ... The area of an exscribed n-gon is

$$ex(n) = n\tan(\pi/n)$$

and that of an inscribed n-gon is

$$in(n) = \frac{n}{2}\sin(2\pi/n).$$

Clearly, for all $n = 4, 5, \ldots,$

$$in(n) < \text{Area of Circle} < ex(n).$$

The two functions are plotted in Figure 11.1 by means of the following *Mathematica* inputs:

```
In[1] := ex[n_]:= n(Tan[Pi/n])
In[2] := in[n_]:= n/2(Sin[2 Pi/n])
In[3] := ListPlot[Table[n/2(Sin[2 Pi/n]) ,{n,4,100}],
                  AxesLabel->{"n",""}];
In[4] := ListPlot[Table[n(Tan[Pi/n]) ,{n,4,100}],
                  AxesLabel->{"n",""}];
In[5] := Show[%,%%];
```

Can you prove that the limits of $ex(n)$ and $in(n)$ both exist and equal π? The fact that you need is

$$\lim_{x \to 0} \frac{\sin x}{x} = 1$$

which is one trigonometric limit that should definitely be commited to memory.

Use the function `Series` to investigate the power series expansions of the trigonometric functions sin, cos, tan near $x = 0$:

```
In[1] := Series[Sin[x],{x,0,10}]
In[2] := Series[Cos[x],{x,0,10}]
In[3] := Series[Tan[x],{x,0,10}]
```

Guess, for each of these, the first two terms of the power series expansion of their inverse. Check your guesses by using `Series`.

11.3. DERIVATIVES AND INTEGRALS OF TRIGONOMETRIC FUNCTIONS

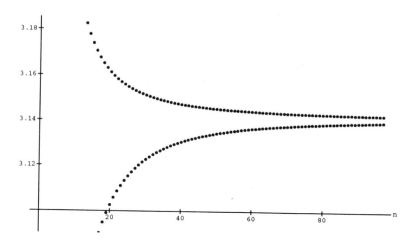

Figure 11.1: The areas of inscribed and exscribed n-gons for a unit circle tend to the limit π

11.3 Derivatives and Integrals of Trigonometric Functions

Let u denote any differentiable function of x. Then the derivatives of the trigonometric functions are as follows:

$$\frac{d}{dx}\sin u = \cos u \frac{d}{dx} u$$
$$\frac{d}{dx}\cos u = -\sin u \frac{d}{dx} u$$
$$\frac{d}{dx}\tan u = \sec^2 u \frac{d}{dx} u$$
$$\frac{d}{dx}\cot u = -\csc^2 u \frac{d}{dx} u$$
$$\frac{d}{dx}\sec u = \sec u \tan u \frac{d}{dx} u$$
$$\frac{d}{dx}\csc u = -\csc u \cot u \frac{d}{dx} u.$$

The Fundamental Theorem of Calculus, and an ability to read in the opposite direction, gives us the integrals of the right hand sides of these equations. The remaining integrals are found by using the definitions

$$\tan x = \frac{\sin x}{\cos x}, \quad \cot x = \frac{1}{\tan x}, \quad \sec x = \frac{1}{\cos x}, \quad \csc x = \frac{1}{\sin x},$$

where, as always, their domains of definition disallow denominators from taking the value zero.

Exercises

1. Plot the trigonometric functions over the interval $[-2\pi, 2\pi]$. Note asymptotes and zeros.

2. For each function, decide where it is increasing, decreasing, and has extreme values. Indicate the interval of values containing 0 for which each function is one-to-one and hence has an inverse.

3. Let $f(x) = a\cos x + b\sin x$ and find the set of vales of a, b for which $\frac{d^2}{dx^2} f = -f$.

4. Use the substitution $w = \cos x$ to evaluate $\int \tan x \, dx$. Try to find a motivation, other than that it works, for the substitution $w = \sec x + \tan x$ in $\int \sec x \, dx$.

5. Complete your own table of integrals and derivatives of the trigonometric functions. By inspection of the graphs, estimate the values of the definite integrals of these functions over the interval $[0, \frac{\pi}{2}]$. Compute the precise values of these definite integrals.

6. Use the identities
$$\sin^2 x = \frac{1}{2}(1 - \cos 2x), \quad \cos^2 x = \frac{1}{2}(1 + \cos 2x)$$
to evaluate $\int \sin^2 x \, dx$ and $\int \cos^2 x \, dx$ Find a similar way to evaluate $\int \tan^2 x \, dx$ and $\int \cot^2 x \, dx$.

7. Find the derivatives of $\csc e^x$, $e^{\cot x}$, $\log|\sin x|$, $\tan^3 \sec^3 x$.

11.4 Inverse Trigonometric Functions

From the above results and exercises we conclude that the trigonometric functions do admit inverses over limited domains. The graph of the inverse of a function is the mirror image, in the line $y = x$, of the graph of the given function and this is the easiest way to sketch it by hand. Formally, the inverse sine function is defined by:
$$\sin^{-1} x = y \text{ if and only if } \sin y = x, \quad \text{valid for } -\frac{\pi}{2} \leq y \leq \frac{\pi}{2}.$$
By differentiating $\sin y = x$ we deduce that
$$\frac{d}{dx} \sin^{-1} x = \frac{1}{\sqrt{1 - x^2}}.$$

Exercises

1. Define the remaining inverse trigonometric functions with their domains, plot them and find their derivatives.

2. Find the derivatives of $\csc^{-1} e^x$, $e^{\cot^{-1} x}$, $\log|\sin^{-1} x|$.

3. Use the substitution $u = x\sqrt{2}$ to evaluate $\int \frac{dx}{x\sqrt{2x^2-1}}$.

11.5 Hyperbolic Functions

Sine and cosine functions satisfy the differential equation

$$\frac{d^2}{dx^2} f = -f$$

while hyperbolic sine and cosine functions satisfy

$$\frac{d^2}{dx^2} f = f.$$

Actually, we are really recycling the exponential function because clearly e^{-x} satisfies the second of these equations just as well as e^x does, and herein lies the definition of the hyperbolic functions:

$$\sinh x = \frac{e^x + e^{-x}}{2}, \quad \cosh x = \frac{e^x - e^{-x}}{2}, \quad \tanh x = \frac{\sinh x}{\cosh x}.$$

Then, analogously to the trigonometric family, we have the reciprocals:

$$\text{cosech} x = \frac{1}{\sinh x}, \quad \text{sech} x = \frac{1}{\cosh x}, \quad \coth x = \frac{1}{\tanh x}$$

with the usual safeguards about division by zero.

Exercises

1. Plot the hyperbolic functions over the interval $[-1, 1]$. Note asymptotes and zeros.

2. For each function, decide where it is increasing, decreasing, and has extreme values. Indicate the set of values in $[-1, 1]$ for which each function is one-to-one and hence has an inverse. What about outside this interval?

3. Verify that $\cosh^2 x - \sinh^2 x = 1$; derive analogous identities to those you know for trigonometric functions and devise a memory aid for the differences.

4. Let $f(x) = a\cosh x + b\sinh x$ and find the set of values of a, b for which $\frac{d^2}{dx^2} f = f$.

5. Compute from the definitions a table of derivatives and integrals of the hyperbolic functions and of their inverse functions.

6. The curve $y = a \cosh \frac{x}{a}$ is called a catenary because it quite well represents the shape taken up by a chain hanging between two supports, catena being Latin for chain. Justify this name by considering the equilibrium of forces on a hanging chain. Find the arc length of a section of the catenary and the surface area of the catenoid obtained by rotating it about the x-axis. Plot the catenoid.

7. The equation $x^2 + y^2 = 1$ for a unit circle can be presented in parametric form in terms of coordinates $x = \cos \theta$, $y = \sin \theta$. The triangles subtended at its centre, are frequently useful in algebraic deductions about trigonometric functions. The corresponding equation for hyperbolic functions is the hyperbola $x^2 - y^2 = 1$. Observe that points on this curve are parametrically given by coordinates $x = \cosh u$, $y = \sinh u$. The angle θ in the circle measures (twice) the area of sector it defines. What area does u measure in relation to $x^2 - y^2 = 1$?

Chapter 12

Techniques of Integration

> *And by these principles, the road lies open for greater things.* —I.Newton, at the end of Tractatus de Quadratura Curvarum 1704

12.1 Symbolic Integration

If $F'(x) = f(x)$, then $\int f(x)dx = F(x) + c$; so every differentiation formula can be written as an indefinite integral. Rewriting basic derivatives in terms of integrals produces the following short list of 17 basic indefinite integrals:

$$\text{(12.1)} \qquad \int x^n dx = \frac{1}{n+1}x^{n+1} + c, \text{ if } n \neq -1$$

$$\text{(12.2)} \qquad \int \frac{1}{x}dx = \log|x| + c$$

$$\text{(12.3)} \qquad \int a^x dx = \frac{a^x}{\log a} + c$$

$$\text{(12.4)} \qquad \int \log x \, dx = x\log x - x + c$$

$$\text{(12.5)} \qquad \int \sin x \, dx = -\cos x + c$$

$$\text{(12.6)} \qquad \int \cos x \, dx = \sin x + c$$

$$\text{(12.7)} \qquad \int \sec^2 x \, dx = \tan x + c$$

$$\text{(12.8)} \qquad \int \csc^2 x \, dx = \cot x \, dx + c$$

$$\text{(12.9)} \qquad \int \sec x \tan x \, dx = \sec x + c$$

$$\text{(12.10)} \qquad \int \csc x \cot x \, dx = -\csc x + c$$

$$(12.11) \quad \int \tan x\, dx = \log|\sec x| + c$$

$$(12.12) \quad \int \cot x\, dx = -\log|\csc x| + c$$

$$(12.13) \quad \int \sec x\, dx = \log|\sec x + \tan x| + c$$

$$(12.14) \quad \int \csc x\, dx = \log|\csc c - \cot x| + c$$

$$(12.15) \quad \int \frac{dx}{\sqrt{a^2 - x^2}} = \arcsin \frac{x}{a} + c$$

$$(12.16) \quad \int \frac{dx}{a^2 + x^2} = \frac{1}{a} \arctan \frac{x}{a} + c$$

$$(12.17) \quad \int \frac{dx}{x\sqrt{x^2 - a^2}} = \frac{1}{a} \operatorname{arcsec} \frac{|x|}{a} + c$$

The problem of solving more complicated integrals in terms of these basic integrals is much more difficult than the corresponding problem of finding derivatives of complicated expressions in terms of derivatives of simpler expressions (by using the product rule, quotient rule and chain rule.) However, over the years, thousands of integration formulas have been developed and tabulated — the inside cover of any calculus text book usually gives a table of about 100 integral formulas.

Since *Mathematica* can integrate many, many symbolic expressions, the need to refer to tables of integrals may well become unnecessary. However, you may wonder how these integral formulas were developed. What are the techniqes used to integrate complicated expressions in terms of simpler ones?

The basic techniques of integration are *algebraic manipulation*, *substitution* or *change of variable*, *parts*, and the method of *partial fractions*. These are standard methods taught in any first year calculus course — and, even though *Mathematica* will probably be able to handle any integration problem you may encounter in a first year calculus course, *you are still responsible for understanding these methods*.

A thorough treatment of these integration methods is offered in Chapter 9 of Edwards and Penney [2]. To help you practice these methods with *Mathematica*, four functions have been defined for you in the package for this chapter. They are **parts, transform, inversesub,** and **completesquare**. Once you have called in the package **Chap12.m**, you can query *Mathematica* about each of these functions. Here are some examples of how to use them.

Example 1. Consider the integral $\int x^2 \cos x^3\, dx$. It can be integrated by using the substitution $u = x^3$. In *Mathematica* we can do it as follows:

```
In[1]:= << Chap12.m
In[2]:= transform[x^2 Sin[x^3], x, u, x^3]
        Sin[u]
        ------
          3
```

12.1. SYMBOLIC INTEGRATION

```
Out[2]= {------, u == x }
          3
```

Whenever you use `transform`, the output will be a list: the first element of which will be the new integrand in terms of the new variable u; the second, an equation giving your substitution, u, in terms of the old variable, x. You should also be aware that `transform` automatically equates the symbol `newintegrand` with the first element of its output list. This allows you to refer to the new integrand, after transformation, via a one-word symbol. This would be particularly useful if the new integrand is a complicated fraction or power — you can save having to type it all out. Note that the new integrand can now be integrated by using one of the 17 basic integration formulas — namely, (12.5).

```
In[3] := Integrate[newintegrand, u]
         -Cos[u]
Out[3] = -------
            3
```

The final step is to replace u by x^3:

```
In[4] := % /. u -> x^3
                 3
         -Cos[x ]
Out[4] = --------
            3
```

Example 2. Now consider the integral $\int x\sqrt{x+3}\,dx$. Try the substitution $u = x+3$:

```
In[5] := transform[x Sqrt[x + 3], x, u, x + 3]
Out[5]= {Sqrt[u] x, u == 3 + x}
```

The new integrand still has a term with x in it, so we need to solve for x in terms of u before we can integrate the new integrand with respect to u. In this case, x can be expressed in terms of u by inspection: $x = u - 3$.

```
In[6] := newintegrand /. x -> u - 3
Out[6]= (-3 + u) Sqrt[u]
```

This can now be expanded and integrated term by term with two applications of (12.1).

```
In[7] := Expand[%]
```

```
Out[7]= -3 Sqrt[u] + u
In[8]:= Integrate[%, u]
                       5/2
            3/2     2 u
Out[8]= -2 u     + ------
                        5
```

Again, the final step is to replace u by $x + 3$:

```
In[9]:= % /. u -> x + 3
                              5/2
              3/2     2 (3 + x)
Out[9]= -2 (3 + x)    + -------------
                              5
```

Example 3. Another approach to the integral of Example 2 is to use what is sometimes referred to as an *inverse substituion* — that is, let x equal an expression in terms of a new variable, say z. Try letting $x = z^2 - 3$, to eliminate the square root in the integrand. You can do this with *Mathematica* by using `inversesub`:

```
In[10]:= inversesub[x Sqrt[x + 3], x, z, z^2 - 3]
                2          2
Out[10]= {2 z  (-3 + z ), {{z -> Sqrt[3 + x]}, {z -> -Sqrt[3 + x]}}}
```

The output is again a list: the first member of which is the new integrand in terms of the new variable, z; the second member, a list of replacement values for z in terms of x. (*Mathematica* may actually generate many more replacement values than actually required — use the one appropriate to the integral you are trying to solve.) This example can now be finished by expansion of the new integrand, two applications of (12.1), and substituting for z:

```
In[11]:= Expand[newintegrand]
              2      4
Out[11]= -6 z  + 2 z
In[12]:= Integrate[%, z]
                        5
              3      2 z
Out[12]= -2 z   + ----
                        5
In[13]:= % /. z -> Sqrt[x + 3]
                              5/2
              3/2     2 (3 + x)
Out[13]= -2 (3 + x)    + -------------
                              5
```

12.1. SYMBOLIC INTEGRATION

A word of warning about using `inversesub` with the trigonometric substitutions $x = a\sin z$, $x = a\tan z$, or $x = a\sec z$: *Mathematica* often has difficulty simplifying trigonometric expressions — it may be necessary to simplify some expressions yourself by explicitly inputting the simpler form. Consider the following example:

Example 4. To integrate $\int \dfrac{dx}{(1+x^2)^{3/2}}$, let $x = \tan z$:

```
In[14]:= inversesub[1/(1 + x^2)^(3/2), x, z, Tan[z]]
Solve::ifun: Warning: inverse functions are being used by Solve,
    so some solutions may not be found.
                       1
Out[14]= {----------------------, {{z -> ArcTan[x]}}}
                 2          2 3/2
             Cos[z]  (1 + Tan[z] )
```

The new integrand is $\dfrac{1}{\cos^2 z (1+\tan^2 z)^{3/2}}$, which *Mathematica* will not simplify, but which you can easily simplify by using

$$1 + \tan^2 z = \sec^2 z.$$

```
In[15]:= newintegrand /. 1 + Tan[z]^2 -> Sec[z]^2
Out[15]= Cos[z]
In[16]:= Integrate[%, z]
Out[16]= Sin[z]
```

Finally, put your answer back in terms of the original variable x:

```
In[17]:= % /. z -> ArcTan[x]
Out[17]= Sin[ArcTan[x]]
```

Mathematica won't simplify this further, but by drawing a triangle and using some elementary trigonometry (how?) you can rewrite this last expression as $\dfrac{x}{\sqrt{x^2+1}}$.

Example 5. The formula for integration by parts is

$$\int u\,dv = uv - \int v\,du.$$

It is a direct consequence of the product rule for derivatives

$$(uv)' = u'v + uv'$$

and the Fundamental Theorem of Calculus. A convenient choice of u and v is governed by the following considerations:

1. The factor u must be differentiated.

2. The factor dv must be integrated for v in terms of x.

3. The newintegral $\int v du$ should be *easier* than the original one.

Consider the example $\int x \cos x \, dx$. This can be integrated by letting $u = x$, $dv = \cos x \, dx$. You can do this by using `parts`. Note that only the choice of u is declared — by elimination the rest of the integrand must be integrated.

```
In[18]:= parts[x Cos[x], x, u, x]
Out[18]= {Sin[x], u -> x, v -> Sin[x]}
```

Note that the output is a list: the first member of which is the new integrand, vdu; the last two members, the replacement values for u and v. As with `transform` and `inversesub`, `parts` automatically equates the symbol `newintegrand` with the new integrand (obtained after applying integration by parts) for future reference. So the problem can be finished by applying the parts formula as follows:

```
In[19]:= ( u v /. %[[{2, 3}]] ) - Integrate[newintegrand, x]
Out[19]= Cos[x] + x Sin[x]
```

Algebraic manipulation is often required to put an integrand into alternate form before one of the basic integral formulas is used, or before a substitution is used, or before integration by parts is used, etc. In one case, if the integrand is a rational function, the usual aglebraic approach is to find the *partial fraction decomposition* of the integrand, and then to integrate each term separately. Solving for the partial fraction decomposition can be a tedious exercise — the good news is that *Mathematica* offers a built-in function, `Apart`, to do it for you. In another case, when a *quadratic expression* occurs in the integrand, you must often complete the square (and then use a trigonometric substitution.) The defined function, `completesquare`, will complete the square for you.

Example 6. Consider the integral $\int \sqrt{2x - x^2} \, dx$. Try `completesquare` on the quadratic expression $2x - x^2$:

```
In[20]:= completesquare[2 x - x^2, x]
                       2
Out[20]= 1 - (-1 + x)
```

which suggests the inverse trigonometric substitution $x = \sin z + 1$.

```
In[21]:= inversesub[Sqrt[2 x - x^2], x, z, Sin[z] + 1]
Solve::ifun: Warning: inverse functions are being used by Solve,
```

12.1. SYMBOLIC INTEGRATION

so some solutions may not be found.

```
Out[21]= {Cos[z] Sqrt[2 (1 + Sin[z]) - (1 + Sin[z])^2 ],

>        {{z -> ArcSin[-1 + x]}}}
In[22]:= Simplify[newintegrand]
Out[22]= Cos[z] Sqrt[1 - Sin[z]^2 ]
```

which can be simplified even further to $\cos^2 z$, and then integrated after using the identity

$$\cos^2 z = \frac{1 + \cos 2z}{2}:$$

```
In[23]:= % /. 1 - Sin[z]^2 -> Cos[z]^2

Out[23]= Cos[z]^2
In[24]:= Integrate[(1 + Cos[2 z])/2, z]
          z    Sin[2 z]
Out[24]= - + --------
          2       4
In[25]:= << Algebra/Trigonometry.m
In[26]:= TrigReduce[Out[24]]
          z    Cos[z] Sin[z]
Out[26]= - + -------------
          2         2
In[27]:= % /. z -> ArcSin[x - 1]
          ArcSin[-1 + x]     (-1 + x) Cos[ArcSin[-1 + x]]
Out[27]= -------------- + ----------------------------
                2                       2
In[28]:= Out[15] /. Cos[ArcSin[-1 + x]] -> Sqrt[2 x - x^2]
          (-1 + x) Sqrt[2 x - x^2 ]    ArcSin[-1 + x]
Out[28]= ----------------------- + --------------
                    2                      2
```

MATHEMATICA FUNCTIONS INTRODUCED IN THIS SECTION
Apart

Exercises:

1. Using the defined functions **parts**, **transform**, **inversesub** and **completesquare**, plus the built-in function **Apart**, integrate each of the fol-

lowing integrals by reducing it to one or more of the basic 17 integral formulas listed at the beginning of this section.

(a) $\int x^2 \sqrt{2x^3 - 4}\,dx$; let $u = 2x^3 - 4$.

(b) $\int e^{\tan x} \sec^2 x\,dx$; let $u = \tan x$.

(c) $\int \frac{1}{\sqrt{x}}(1+\sqrt{x})^4\,dx$; let $u = 1 + \sqrt{x}$.

(d) $\int \frac{e^{2x}}{1+e^{4x}}\,dx$; let $u = e^{2x}$.

(e) $\int \frac{1}{\sqrt{e^{2x} - 1}}\,dx$; let $u = e^x$.

(f) $\int \sin^3 x\,dx$; let $u = \cos x$.

(g) $\int \frac{\sin^3 x}{\sqrt{\cos x}}\,dx$; let $u = \cos x$.

(h) $\int \tan x \sec^4 x\,dx$; let $u = \tan x$.

(i) $\int \frac{\tan x + \sin x}{\sec x}\,dx$; let $u = \sin x$.

(j) $\int x e^{-x}\,dx$; use parts with $u = x$.

(k) $\int x^2 e^{-x}\,dx$; use parts with $u = x^2$.

(l) $\int x^2 \log x\,dx$; use parts with $u = \log x$.

(m) $\int \arctan x\,dx$; use parts with $u = \arctan x$.

(n) $\int \arctan \sqrt{x}\,dx$; use parts with $u = \arctan \sqrt{x}$.

(o) $\int e^{ax} \sin bx\,dx$; use parts with $u = e^{ax}$.

(p) $\int \frac{1}{\sqrt{x}(1+x)}\,dx$; let $x = z^2$.

(q) $\int \frac{1}{x^2\sqrt{a^2 - x^2}}\,dx$; let $x = a \sin z$.

(r) $\int \frac{x^2}{\sqrt{a^2 + x^2}}\,dx$; let $x = a \tan z$.

(s) $\int \frac{\sqrt{x^2 - 25}}{x}\,dx$; let $x = 5 \sec z$.

(t) $\int \frac{x^3 - 1}{x^3 + x}\,dx$; use partial fractions.

(u) $\int \dfrac{x^2 - 10}{2x^4 + 9x^2 + 4} dx$; use partial fractions.

(v) $\int \dfrac{5 - 3x}{x^2 + 4x + 5} dx$; complete the square first.

2. Explain why you think *Mathematica* gave the warning about inverse functions in Out[21]; give an example.

3. Describe in your own words how you think transform, inversesub, and parts work. Could a machine follow the steps you describe? Look at the definitions of transform, inversesub, and parts and compare them with your steps. Are they the same?

12.2 The Integration Game

OBJECT OF THE GAME: To reduce the integration of a given integrand to one or more applications of the basic integration formulas, (12.1) to (12.17), in as few steps as possible.

RULES OF THE GAME: At any step you can use one or more of the defined functions parts, transform, inversesub, completesquare; one or more of *Mathematica's* built in functions for algebraic manipulation — Expand, Simplify, Factor, Apart; or one or more replacements, via /. *expression -> different expression*. The only real restriction in this game is that you can use Integrate only if the integral is a sum of, or a constant multiple of, (or a sum of multiples of), one or more of the formulas (12.1) to (12.17).

HOW TO PLAY:

1. To establish a reference point in your session, take the given integral and Integrate it at once to find *Mathematica's* one–line answer. You may, if you wish, Expand or Simplify this output for future reference; call it Out[m].

2. Use the Rules of the Game on the given integral until you have solved it. Your answer should be presented as one output; call it Out[n].

3. You are finished playing if *Mathematica's* response to Out[m] == Out[n] is True.

4. You win if your score, $n - m$, is less than your competitor's.

To play the game, you must have access to the package Chap12.m, so let us call it in right now:

In[1] := << Chap12.m

Example. Consider the integral

$$\int \frac{x^3}{\sqrt{1-x^2}}\,dx.$$

Here is *Mathematica's* one-line answer:

```
In[2]:= Integrate[x^3 / Sqrt[1 - x^2], x]
                                   2 3/2
                       2    2 (1 - x )
           -2 Sqrt[1 - x ] + --------------
                                    3
Out[2]= -------------------------------
                       2
In[3]:= Expand[%]    (* to get rid of common factor 2 *)
                               2 3/2
                     2    (1 - x )
Out[3]= -Sqrt[1 - x ] + -----------
                             3
```

Player 1: Plays it by the book; starts off with the trigonometric substitution $x = \sin z$:

```
In[4]:= inversesub[x^3 / Sqrt[1 - x^2], x, z, Sin[z]]
Solve::ifun: Warning: inverse functions are being used by Solve, so some
    solutions may not be found.
                         3
             Cos[z] Sin[z]
Out[4]= {-------------------, {{z -> ArcSin[x]}}}
                          2
             Sqrt[1 - Sin[z] ]
In[5]:= newintegrand /. 1 - Sin[z]^2 -> Cos[z]^2
               3
Out[5]= Sin[z]
In[6]:= transform[%, z, w, Cos[z]]
                     2
Out[6]= {-Sin[z]  , w == Cos[z]}
In[7]:= newintegrand /. Sin[z]^2 -> 1 - w^2
                 2
Out[7]= -(1 - w )
In[8]:= Integrate[%, w]  (* two applications of (12.1) *)
                 3
                w
Out[8]= -w + --
```

12.2. THE INTEGRATION GAME

```
                             3
In[9]:= % /. w -> Cos[z]
                              3
                         Cos[z]
Out[9]= -Cos[z] + -------
                         3
In[10]:= % /. z -> ArcSin[x]
                                      3
                         Cos[ArcSin[x]]
Out[10]= -Cos[ArcSin[x]] + ---------------
                                   3
In[11]:= % /. Cos[ArcSin[x]] -> Sqrt[1 - x^2]
                           2 3/2
                  2     (1 - x )
Out[11]= -Sqrt[1 - x ] + -----------
                              3
In[12]:= Out[11] == Out[3]
Out[12]= True
```

So Player 1's score is 8.

Player 2: Hopes to avoid trigonometry whenever possible; starts with a simple substitution: let $u = 1 - x^2$.

```
In[4]:= transform[x^3 / Sqrt[1 - x^2], x, u, 1 - x^2]
              2
            -x                  2
Out[4]= {---------, u == 1 - x }
          2 Sqrt[u]
In[5]:= newintegrand /. x^2 -> 1 - u
         -(1 - u)
Out[5]= ---------
         2 Sqrt[u]
In[6]:= Expand[%]
            -1        Sqrt[u]
Out[6]= --------- + -------
         2 Sqrt[u]     2
In[7]:= Integrate[%, u]   (* two applications of (12.1) *)
                    3/2
                   u
Out[7]= -Sqrt[u] + ----
                    3
In[8]:= % /. u -> 1 - x^2
                       2 3/2
```

```
                                 2        (1 - x )
Out[8]= -Sqrt[1 - x ] + -----------
                                          3
In[9]:= Out[8] == Out[3]
Out[9]= True
```

So Player 2's score is only 5.

Player 3: Tries to be creative; starts off with the inverse substitution $z^2 = 1 - x^2$.

```
In[4]:= inversesub[x^3 / Sqrt[1 - x^2], x, z, Sqrt[1 - z^2]]
                    2
            z (1 - z )                                    2                     2
Out[4]= {-(------------------),{{z->Sqrt[1- x ]},{z->-Sqrt[1 - x]}}}
                    2
          Sqrt[1 - (1 - z )]
In[5]:= Simplify[newintegrand]
                  2
Out[5]= -1 + z
In[6]:= Integrate[%, z]   (* two applications of (12.1) *)
              3
             z
Out[6]= -z + --
             3
In[7]:= % /. z -> Sqrt[1 - x^2]
                              2 3/2
                    2    (1 - x )
Out[7]= -Sqrt[1 - x ] + -----------
                              3
In[8]:= Out[7] == Out[3]
Out[8]= True
```

So Player 3's score is only 4!

Player 4: Remembers only one thing from high school calculus — integration by parts; tries parts on *every* integral. Let $u = x^2$, let $dv = \int \frac{x}{\sqrt{1-x^2}} dx$.

```
In[4]:= parts[x^3 / Sqrt[1 - x^2], x, u, x^2]
                        2          2                       2
Out[4]= {-2 x Sqrt[1 - x ], u -> x , v -> -Sqrt[1 - x ]}
In[5]:= transform[newintegrand, x, a, 1 - x^2]
                                2
Out[5]= {Sqrt[a], a == 1 - x }
In[6]:= Integrate[newintegrand, a] (* one application of (12.1) *)
```

12.2. THE INTEGRATION GAME

```
                3/2
             2 a
Out[6]=     ------
               3
In[7]:= % /. a -> 1 - x^2
                   2 3/2
             2 (1 - x )
Out[7]=     -------------
                  3

In[8]:= ( u v /. Out[4][[{2, 3}]] ) - %    (* parts formula *)
                                     2 3/2
             2              2   2 (1 - x )
Out[8]= -(x   Sqrt[1 - x  ]) - -------------
                                     3
```

Although this is a perfectly acceptable answer, *Mathematica* does not respond with True if you enter Out[8] == Out[3].

```
In[9]:= Out[8] == Out[3]
                                2 3/2                          2 3/2
      2              2    2 (1 - x )              2     (1 - x )
 -(x   Sqrt[1 - x  ]) - ------------- == -Sqrt[1 - x  ] + -----------
                              3                                3
```

Thus more manipulation is required; either way Player 4's score will be greater than 5 — and Player 3 is the winner.

NOTE: Actually each of the above 4 players played very well in the sense that none of them made typing errors or syntax errors (which would waste steps) and every choice made was actually a step forward — no steps were totally useless. Beginners at the Integration Game will not always be so lucky — it takes a lot of experience with the techniques of integration before you can play the Integration Game well!

FUNCTIONS CONTAINED IN THE PACKAGE Chap12.m
transform   inversesub   parts
completesquare

Exercises: Play the Integration Game with the following integrals; a good score is indicated to the right of each integral.

1. $\int \frac{1}{(1+x^2)^2} dx$; 5 steps

2. $\int \dfrac{x^3}{x^2 + a^2} dx$; 4 steps

3. $\int \dfrac{x^{7/2}}{1 + \sqrt{x}} dx$; 4 steps

4. $\int \dfrac{xe^x}{(1+x)^2} dx$; 4 steps

5. $\int \dfrac{1}{\sqrt{x^2 - 4x + 13}} dx$; 8 steps

6. $\int x \arcsin x\, dx$; 7 steps

7. $\int \dfrac{dx}{\sqrt{\tan x}}$; 12 steps (a challenge!)

Chapter 13

Calculus Projects

Recipe for chaos. —I.N.Stewart, Does God play dice?, 1989

This chapter provides some directions which you may find interesting to pursue.

13.1 Investigating Chaos

Section 4.4 was only a brief introduction to the method of fixed point iteration. That method could be used sometimes to find a fixed point of a function. There is a very powerful result, called the Brouwer Fixed Point Theorem, which says that every continuous function from a closed interval to itself *has* at least one fixed point. Indeed this result is also true for self-maps of products of intervals (squares, cubes, hypercubes of any dimension) and their equivalents like disks, solid spheres, etc. So, in three dimensions, if you stir a cup of coffee you may conclude that at least one of its constituent particles is in its original position before stirring!

The Brouwer Fixed Point Theorem is very widely used in analysis and can be proved using algebraic topological methods which are generalizations of those used to prove the Intermediate Value Theorem. However, these are *existence* proofs and do not tell us how to *find* the fixed points. When the function in question is differentiable then we can obtain constructive methods to approach a fixed point iteratively. This section states and discusses a result which can be used to determine if functional iteration converges to a fixed point of a function g. We investigate briefly what may happen if functional iteration does not converge and then invite you to pursue the matter further. References for further reading are listed at the end of this section.

Theorem 13.1 (Fixed Point Theorem) *Suppose g is a continuous function such that*
$$g : [a,b] \longrightarrow [a,b].$$

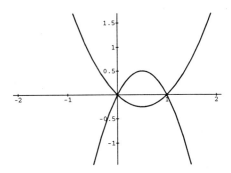

Figure 13.1: $y = kx(1-x)$; if $k > 0$ the parabola opens downwards; if $k < 0$ the parabola opens upwards.

If there is a number $r < 1$ such that $|g'(x)| \leq r$ for all x in (a,b), then there is a unique fixed point p of g in the interval $[a,b]$, and the sequence

$$x_n = g(x_{n-1}), \text{ for } n \geq 1$$

converges to p for any choice of x_0 in $[a,b]$.

You can use calculus to prove this Fixed Point Theorem: the existence of a fixed point is a consequence of the Intermediate Value Theorem; the uniqueness of the fixed point and the convergence of the sequence x_n are consequences of the Mean Value Theorem.

If p is a fixed point of g and $|g'(p)| < 1$, then p is called a *stable equilibrium value* of g, since in this case functional iteration applied to an initial choice, x_0, close enough to p will result in a sequence converging to p. However, if p is a fixed point of g and $|g'(p)| > 1$, then p is called an *unstable equilibrium value* of g, since in this case functional iteration applied to any initial choice (not equal to p) will result in a sequence that *does not* converge to p.

To illustrate all of the above, consider the family of parabolas with equation

$$y = kx(1-x).$$

If $k > 0$ then the parabola opens downwards; if $k < 0$, the parabola opens upwards. In each case the graph has two x–intercepts, 0 and 1. See Figure 13.1

```
In[1]:= g[k_][x_] := k x (1 - x)
In[2]:= Plot[{g[2][x], g[-1][x]}, {x, -1, 2}, AspectRatio -> Automatic];
```

13.1. INVESTIGATING CHAOS

We can use *Mathematica* to find the fixed points of $g[k]$, and the derivatives at each of the (two) fixed points:

```
In[3]:= Solve[g[k][x] == x, x]
Out[3]= {{x -> 1 - 1/k}, {x -> 0}}
In[4]:= g[k]'[x] // Simplify
Out[4]= k - 2 k x
In[5]:= {g[k]'[0], g[k]'[1 - 1/k]} // Simplify
Out[5]= {k, 2 - k}
```

You can check that for $0 < k < 4$ we have

$$g[k] : [0, 1] \longrightarrow [0, 1].$$

Thus by the Fixed Point Theorem, 0 is a stable equilibrium value of $g[k]$ but $1 - 1/k$ is an unstable equilibrium value of $g[k]$, if $0 < k < 1$. On the other hand, if $1 < k < 3$ then 0 is the unstable equilibrium value and $1 - 1/k$ is the stable equilibrium value. We can illustrate the above with calculations by making use of **FixedPoint**:

```
In[6]:= FixedPoint[g[0.5], 0.1]
Out[6]= 0.
In[7]:= FixedPoint[g[1.5], 0.1]    (* 1 - 1/k = 1/3 *)
Out[7]= 0.333333
In[8]:= FixedPoint[g[2.6], 0.1]    (* 1 - 1/k = 8/13 *)
Out[8]= 0.615385
```

If we call in the package **Chap4.m** we can use the defined function **fixedpoint** to see what is happening.

```
In[9]:= << Chap4.m  (* Out[9] is a usage statement *)
```

If $k = 2.6$ we see that functional iteration quickly converges to the fixed point of 8/13. See Figure 13.2.

```
In[10]:= fixedpoint[g[2.6], 0.1, {0, 1}][30];
```

We now proceed to investigate what happens if $k > 3$. If $k = 3.2$ then

$$g[k]'(1 - 1/k) > 1,$$

the equilibrium value is unstable and no convergence occurs. See Figure 13.2.

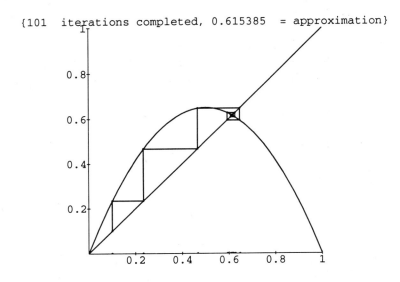

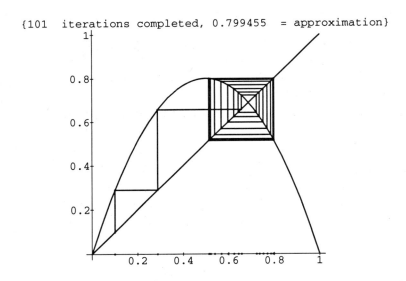

Figure 13.2: Functional iteration applied to $y = kx(1-x)$ with initial value 0.1. If $k = 2.6$ convergence to the stable equilibrium value of 0.615385 occurs (top); if $k = 3.2$ no convergence occurs after 101 iterations (bottom).

13.1. INVESTIGATING CHAOS

```
In[11]:= fixedpoint[g[3.2], 0.1, {0, 1}][100];
```

Although it is not completely clear from Figure 13.2, what appears to be happening is that the values of $g[3.2]$ oscillate between two different values, neither of which is the fixed point $1 - 1/k = 11/16$. We can check our suspicions by using `FixedPoint`:

```
In[12]:= FixedPoint[g[3.2], 0.1, 101]
Out[12]= 0.799455
In[13]:= FixedPoint[g[3.2], 0.1, 102]
Out[13]= 0.513045
In[14]:= FixedPoint[g[3.2], 0.1, 103]
Out[14]= 0.799455
In[15]:= FixedPoint[g[3.2], 0.1, 104]
Out[15]= 0.513045
```

Indeed it seems as if the successive iterations of $g[3.2]$ are trapped in a periodic cycle of two numbers 0.799455 and 0.513045. If $k = 4$ things are even more interesting. See Figure 13.3

```
In[16]:= fixedpoint[g[4], 0.1, 0, 1][100];
```

In this case, there is no discernible pattern at all — succesive iterations seem totally random. What are we to make of this? For that matter, can we even explain what happened for $k = 3.2$? We can gain some insight into the situation for $k = 3.2$ if we consider the function $g[3.2] \circ g[3.2]$. If we plot it, see Figure 13.4, and consider its equilibrium values things become somewhat clearer.

```
In[17]:= Plot[Nest[g[3.2], x, 2],x,0,1,AspectRatio -> Automatic];
```

What we see is that $g[3.2] \circ g[3.2]$ has four equilibrium values, but only two of them are stable. The unstable equilibrium values of $g[3.2]$, namely 0 and 0.6875 are also unstable equilibrium values of $g[3.2] \circ g[3.2]$. Successive iterations of $g[3.2]$ then alternate between the two stable fixed points of $g[3.2] \circ g[3.2]$. If we solve for these two stable fixed points we find exactly the numbers in Out[12] and Out[13].

```
In[18]:= Solve[Nest[g[16/5], x, 2] == x, x]
                       11         336 + 16 Sqrt[21]
Out[18]= {{x -> 0}, {x -> --}, {x -> ------------------},
                       16               512

              336 - 16 Sqrt[21]
 >     {x -> ------------------}}
                    512
In[19]:= N[%]
Out[19]= {{x -> 0.}, {x -> 0.6875}, {x -> 0.799455}, {x -> 0.513045}}
```

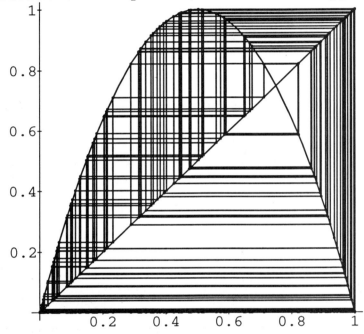

Figure 13.3: If $y = 4x(1 - x)$, functional iteration (applied to the initial choice of 0.1) exhibits no clear pattern.

13.1. INVESTIGATING CHAOS   255

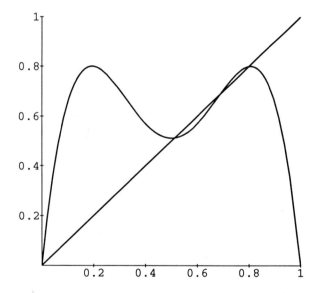

Figure 13.4: For the graph of $g[3.2] \circ g[3.2]$, the two equilibrium values of $g[3.2]$ are both unstable. Successive iterations of $g[3.2]$ alternate between the two stable equilibrium values of $g[3.2] \circ g[3.2]$, namely 0.513045 and 0.799455.

You should now try the exercises for this section to pursue these investigations further. You may also wish to refer to the following books.

For further information:

- P. Davies. **The Cosmic Blueprint** Simon & Schuster, New York 1988

- I. Ekeland. **Mathematics and the Unexpected** University of Chicago Press, Chicago 1988

- J. Gleick. **Chaos** Viking Penguin Inc., New York 1987

- T.W. Gray. and J. Glynn. **Exploring Mathematics with** *Mathematica* Addison–Wesley, California 1991

- H-O. Peitgen. and P.H. Richter. **The Beauty of Fractals** Springer–Verlag, Berlin 1986

- I. Stewart. **Does God Play Dice?** Penguin Books, Markham, Ont. 1990

Exercises:

1. Explain graphically (or prove analytically, if you wish) why functional iteration will converge to a fixed point p if $|g'(p)| < 1$, but will not converge to the fixed point p if $|g'(p)| > 1$.

2. Show that for $0 < k < 4$,
$$g[k] : [0,1] \longrightarrow [0,1].$$

3. Investigate the fixed points of $g[1]$ and $g[3]$. Will functional iteration converge to the fixed points 0 and 2/3, respectively?

4. Another way to see the results of functional iteration is to use `ListPlot` to graph successive iterations. Try this for the cases $k = 2.6$ and $k = 3.2$, `In[10]` and `In[11]` above, as follows:

    (a) `ListPlot[Table[{x, FixedPoint[g[2.6], 0.1, x]}, {x, 1, 100}]]`

    (b) `ListPlot[Table[{x, FixedPoint[g[3.2], 0.1, x]}, {x, 1, 100}]]`

    The first graph clearly shows that functional iteration converges to a fixed point if $k = 2.6$; the second clearly shows that for $k = 3.2$ functional iteration does not converge to a fixed *point* but rather to a cycle of period 2.

5. To investigate things further you should define the following function:

13.1. INVESTIGATING CHAOS

```
iterations[k_, a_, n_] := Block[{x},
  ListPlot[Table[{x, FixedPoint[g[k], a, x]}, {x, 1, n}]]]
```

This will allow you to see the result of n iterations of $g[k]$ applied to the initial choice of $x = a$. We will not be interested in the initial choice $x = a$ as much as we will be in the value of k. Of course, $g[k]$ must be defined as previously:

```
g[k_][x_] := k x (1 - x).
```

For most purposes $n = 100$ iterations will suffice; but sometimes you will need to raise n to 200 or even 300. The larger n the longer you will have to wait for *Mathematica* to create the plot.

6. Use `iterations` with the following values of k to find the length of the periodic cycle, if any. Use `FixedPoint` to find the values of the points in the cycles.

    (a) $k = 3.5$
    (b) $k = 3.561$
    (c) $k = 3.5667$
    (d) $k = 3.6$

7. The value of k at which a cycle doubles in length is called a *bifurcation* point. So $k = 3$ is the bifurcation point at which the length of the periodic cycle changes from 1 to 2. What are the values of the bifurcation points at which the length of the periodic cycle changes from

    (a) 2 to 4?
    (b) 4 to 8?
    (c) 8 to 16?

8. The period doubling of the previous two exercises keeps on occurring to produce cycles of 16, 32, 64, 128, ... but the corresponding bifurcation points are closer and closer together. How would you describe what has happened for $k = 3.6$? Is there a periodic cycle at all?

9. When there is no periodic cycle, the result of functional iteration is described as *chaotic*. In light of the previous three exercises, you might think that once k has passed 3.6, functional iteration produces only chaos. Check this for the following values of k. What do you notice?

    (a) $k = 3.7$
    (b) $k = 3.835$
    (c) $k = 3.739$

(d) $k = 4$

10. Find a value of k slightly larger than 3.835 which results in a periodic cycle of length 6.

11. Find a value of k slightly larger than 3.739 which results in a periodic cycle of length 10.

12. For different values of k, repeated functional iteration of $g[k]$ results in a periodic cycle, or in chaos. It is very revealing to plot the periodic cycles (or the chaotic range of values) against the parameter k. The resulting figure is displayed in Peitgen and Richter, Figure 21, p 25, or in Stewart, Figure 65, p 162. This exercise will attempt to create some of that figure by using simple *Mathematica* functions. (For a more detailed approach, see Gray and Glynn, Chapter 7.)

    (a) Define a function section as follows:

    ```
    section[k_] := Block[{n},
    ListPlot[Table[{k, FixedPoint[g[k], 0.1, n}, {n, 300, 350}]]]
    ```

    Section[k] plots the last 50 results of iterating $g[k]$ 350 times. The hope is that if functional iteration results in a periodic cycle, 350 iterations are enough to exhibit the cycle. Otherwise, 50 values with no discernible pattern will be displayed. (Each plot section[k] is a cross–section of the graph we are trying to create.)

    (b) Plot section[k] for the following values of k:

    i. $k = 2.5$
    ii. $k = 2.8$
    iii. $k = 2.9$
    iv. $k = 3$
    v. $k = 3.1$
    vi. $k = 3.3$
    vii. $k = 3.5$
    viii. $k = 3.561$
    ix. $k = 3.5667$
    x. $k = 3.6$
    xi. $k = 3.7$
    xii. $k = 3.739$
    xiii. $k = 3.742$
    xiv. $k = 3.8$
    xv. $k = 3.82$

xvi. $k = 3.835$
xvii. $k = 3.846$
xviii. $k = 3.9$
xix. $k = 3.95$
xx. $k = 4$

(c) Combine the above 20 graphs with **Show**; you should obtain a sketchy version of Figure 21, p 25 of Peitgen and Richter.

(d) By filling in more and more cross–sections you can improve the details of your graph.

13.2 Newton's Method in the Complex Plane

Newton's method finds approximate solutions x_n to the equation $f(x) = 0$ by means of the iteration formula

$$(13.1) \qquad x_{n+1} = x_n - \frac{f(x_n)}{f'(x_n)}, \text{ for } n \geq 0,$$

with x_0 being the initial choice. This section will investigate how the choice of x_0 determines to which root of $f(x) = 0$, if any, the sequence x_n converges. Recall that the formula for Equation 13.1 is defined as **iteration[f]** in the package **Chap4.m**. We shall call this package in, and then proceed to investigate the convergence of Newton's method for the two equations

$$x^3 - 3x^2 - x + 3 = 0 \text{ and } x^4 - 1 = 0,$$

which have roots ± 1, 3 and ± 1, $\pm i$, respectively.

Example 1:

```
In[1]:= << Chap4.m
                            (* Out[1] will be a usage statement *)
In[2]:= f[x_] := x^3 - 3 x^2 - x + 3
In[3]:= Solve[f[x] == 0, x]
Out[3]= {{x -> -1}, {x -> 3}, {x -> 1}}
```

By making use of **iteration[f]**, define a function **root**, as follows, to perform n iterations of Newton's method with initial choice x:

```
In[4]:= root[f_, x_, n_] := Nest[iteration[f], N[x], n]
```

In this function, the number of iterations can be controlled. If convergence is very "slow", then a large value of n will be necessary to obtain an actual root of $f(x) = 0$. If there is no convergence, then it does not matter how large n is — you will not obtain a root. In these examples, we will usually take $n = 20$ or 30. If you pursue your own examples, you may find it necessary to take different values of n. To begin with, pick initial choices 4.5, −4.5, and 0.3. In each case, after 20 iterations, Newton's method converges to the nearest root:

```
In[5] := root[f, 4.5, 20]
Out[5]= 3.
In[6] := root[f, -4.5, 20]
Out[6]= -1.
In[7] := root[f, 0.3, 20]
Out[7]= 1.
```

However, Newton's method does not alway converge to the nearest root. Consider an initial choice of 0, which is equally near the two roots of ±1. In this case, we find that Newton's method converges to 3, which is the root *furthest* away from the initial choice.

```
In[8] := root[f, 0, 20]
Out[8]= 3.
```

Let us now construct a table for 61 different initial choices, ranging from −2 to 4, in stepsizes of 0.1:

```
In[9] := Table[{i, root[f, i, 20]}, {i, -2, 4, 0.1}]
Out[9]= {{-2, -1.},{-1.9, -1.},{-1.8, -1.},{-1.7, -1.},{-1.6, -1.},

   {-1.5, -1.}, {-1.4, -1.}, {-1.3, -1.}, {-1.2, -1.}, {-1.1, -1.},
   {-1., -1.}, {-0.9, -1.}, {-0.8, -1.}, {-0.7, -1.}, {-0.6, -1.},
   {-0.5, -1.}, {-0.4, -1.}, {-0.3, -1.}, {-0.2, -1.}, {-0.1, 3.},
   {0.0 , 3.}, {0.1, 3.}, {0.2, 1.}, {0.3, 1.}, {0.4, 1.},
   {0.5, 1.}, {0.6, 1.}, {0.7, 1.}, {0.8, 1.}, {0.9, 1.}, {1., 1.},
   {1.1, 1.}, {1.2, 1.}, {1.3, 1.}, {1.4, 1.}, {1.5, 1.}, {1.6, 1.},
   {1.7, 1.}, {1.8, 1.}, {1.9, -1.}, {2., -1.}, {2.1, -1.}, {2.2, 3.},
   {2.3, 3.}, {2.4, 3.}, {2.5, 3.}, {2.6, 3.}, {2.7, 3.}, {2.8, 3.},
   {2.9, 3.}, {3., 3.}, {3.1, 3.}, {3.2, 3.}, {3.3, 3.}, {3.4, 3.},
   {3.5, 3.}, {3.6, 3.}, {3.7, 3.}, {3.8, 3.}, {3.9, 3.}, {4., 3.}}
```

Note that mostly Newton's method does converge to the root nearest the initial choice, but there are exceptions, namely if the initial choice is −0.1, 0.0 or 0.1, or if the initial choice is 1.9, 2.0 or 2.1.

13.2. NEWTON'S METHOD IN THE COMPLEX PLANE

Example 2: Let us now consider the equation $x^4 - 1 = 0$, which has two real roots, and two imaginary roots.

```
In[10]:= f[x_] := x^4 - 1
In[11]:= Solve[f[x] == 0, x]
Out[11]= {{x -> 1}, {x -> I}, {x -> -1}, {x -> -I}}
```

Similar to what we did in Example 1, we can create a table showing to which root Newton's method converges, for initial choices between -2 and 2:

```
In[12]:= Table[{i, root[f, i, 20]}, {i, -2, 2, 0.1}]
Out[12]= {{-2, -1.},{-1.9, -1.},{-1.8, -1.},{-1.7, -1.},{-1.6, -1.},
  {-1.5, -1.}, {-1.4, -1.}, {-1.3, -1.}, {-1.2, -1.}, {-1.1, -1.},
  {-1., -1.}, {-0.9, -1.}, {-0.8, -1.}, {-0.7, -1.}, {-0.6, -1.},
  {-0.5, -1.}, {-0.4, -1.}, {-0.3, -1.}, {-0.2, -1.}, {-0.1, -1.},

 {6.38378 10^-16, 2.28814 10^41}, {0.1, 1.}, {0.2, 1.}, {0.3, 1.},
  {0.4, 1.}, {0.5, 1.}, {0.6, 1.}, {0.7, 1.}, {0.8, 1.}, {0.9, 1.},
  {1., 1.}, {1.1, 1.}, {1.2, 1.}, {1.3, 1.}, {1.4, 1.}, {1.5, 1.},
  {1.6, 1.}, {1.7, 1.}, {1.8, 1.}, {1.9, 1.}, {2., 1.}}
```

Except for the singularity at the initial choice of 0 (which is not unexpected since $f'(0) = 0$), in each case of the above table Newton's method converges to the nearest root. However, things do not stay so predictable if we allow complex numbers as initial choices.[1]

To begin with, let us take an initial choice of $1 + i$, where i is the square root of -1. Newton's method then converges to the root i:

```
In[13]:= root[f, 1 + 2 I, 20]
Out[13]= 1. I
```

Instead of investigating things further with tables of calculations, let us use colours.[2] To do this we shall represent the complex number $x + yi$ by the point (x, y) in the plane. If the initial choice $x + yi$ results in Newton's method converging to the root 1 we shall colour in the point (x, y) red; if the convergence is to the root i we shall colour the point (x, y) green. Similarily we shall use blue to colour in initial points which result in convergence to -1, and yellow for initial points which result in convergence to $-i$. If there is no convergence, we shall colour in the initial point magenta.

---

[1] We shall not cover any calculus of complex numbers here; we only point out that all the calculus required in Newton's method works equally well for complex numbers.

[2] You need a colour monitor for the rest of this section.

There is a package, `Newtons.m`, which contains the necessary functions to do all of the above colouring. In particular, all you need is the function `patch`. Explicitly, `patch[f, {a, b}, {c, d}, n]` determines to which root, if any, Newton's method converges after n iterations, for the initial choices $x + yi$, where x ranges from a to b, and y ranges from c to d, with stepsize 1/10 of the interval length in each direction; and then colours in the 121 points (x, y) accordingly.

```
In[14]:= << Newtons.m
In[15]:= patch[f, {1, 2}, {1, 2}, 30]
Out[15]= -Graphics-
```

Note that the only initial choices which do not result in convergence are all along the line $y = x$, that among the other initial choices there is convergence to each of the four roots, and that the pattern of colours is quite intricate. You can now work through the following exercises to pursue these investigations further. In addition, you may wish to refer to the following books.

For further information:

- M. Barnsley. **Fractals Everywhere** Academic Press, Inc., San Diego, CA, 1988.

- I. Ekeland. **Mathematics and the Unexpected** University of Chicago Press, Chicago 1988.

- J. Gleick. **Chaos** Viking Penguin Inc., New York 1987

- H-O. Peitgen. and P.H. Richter. **The Beauty of Fractals** Springer–Verlag, Berlin 1986.

- I.N. Stewart. **Does God Play Dice?** Penguin Books, Markham, Ont. 1990.

Exercises:

1. Let $f(x) = x^3 - 3x^2 - x + 3$. Construct tables of data as in Example 1, but with $n = 30$, to investigate the convergence of Newton's method on the following intervals:

    (a) $[-0.2, -0.1]$, with stepsize 0.01
    (b) $[-0.16, -0.15]$, with stepsize 0.001
    (c) $[-0.155, -0.154]$, with stepsize 0.0001

2. Repeat Exercise 1 for f on the following intervals:

    (a) $[0.1, 0.2]$, with stepsize 0.01

(b) $[0.1, 0.11]$, with stepsize 0.001

(c) $[0.104, 0.106]$, with stepsize 0.0001

3. What is interesting about the results of Exercise 2 in comparison to the results of Exercise 1? Does the question of which root Newton's method converges to, for a given initial guess, seem to get simpler or more complicated as we focus in on smaller and smaller intervals?

4. Now let $f(x) = x^4 - 1$. Use patch as in Example 2 for the unit square described by $\{a, b\} = \{-2, -1\}$ and $\{c, d\} = \{-2, -1\}$.

5. Repeat the previous exercise for the 15 other unit squares that make up the square bounded by the four corner points

$$(-2, -2), \quad (2, -2), \quad (2, 2), \quad (-2, 2).$$

Use Show to combine all of the previous 16 "patches" into one coloured "mosaic." Comment on the pattern: Does it have any symmetry? Is it possible to predict which root Newton's method will converge to for a given initial point?

6. Let $f(x) = x^4 - 1$. Use patch as in Example 2, but now set $\{a, b\} = \{1.5, 1.6\}$ and $\{c, d\} = \{1.5, 1.6\}$, and take $n = 50$, to get convergence at every point off the line $y = x$. Note the isolated "green" point corresponding to the initial choice of $1.57 + 1.51i$. Investigate what is happening in the vicinity of this point by using patch with $\{a, b\} = \{1.565, 1.575\}$, $\{c, d\} = \{1.505, 1.515\}$, and $n = 50$. Does the pattern simplify, or are there still isolated points?

7. If you wish to use patch for other functions, you need to define the new function, f, but you also need to re-define the function colour which is in the package Newtons.m. The Which statement in the definition of colour states how many roots f has, and which colour should be used depending on which root Newton's method converges to. By making the appropriate changes, use patch to investigate the convergence of Newton's method for different initial choices and the equation $x^3 - 1 = 0$.

13.3 Modelling

In this section we present a few ideas for investigating mathematical models of two rather different real processes.

Growth

We know that pure exponential functions model processes growing in proportion to their size at each instant: $f' = kf$, has exponential solutions. Consider now the

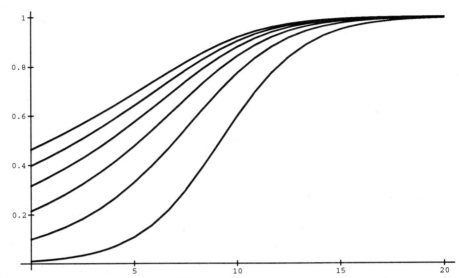

Figure 13.5: The function $(1 + me^{-kt})^{-1/n}$ for $k = 1/2$, $m = 1$, $n = 1, 2, 3, 4, 5, 6$.

family of related functions

$$f(n, t) = (1 + me^{-kt})^{-1/n}, \quad \text{for } n = 1, 2, \ldots, \ k, m > 0.$$

For example, Figure 13.5 shows their graphs for $k = 1/2$, $m = 1$, $n = 1, 2, 3, 4, 5, 6$. These functions are used in botanical research to model the growth of plants—in one study, the blue agave cactus like the one from which tequila is made. However, the characteristic S shape is encountered in many phenomena from economics to engineering and chemistry. We get the graph by the following inputs:

```
In[1]:= f[n_,t_]:=(1+100 E^(-t/2))^(-1/n)
In[2]:= Plot[{f[1,t],f[2,t],f[3,t],f[4,t],f[5,t],f[6,t]},{t,0,20}];
```

You can try some other families and attempt to fit any growth data that you can obtain.

Investigate the differential equation that f must satisfy in t for fixed k, m, n. Try taking the logarithm of f before differentiating, and hence eliminate the explicit appearance of t. Is f always increasing?

13.3. MODELLING

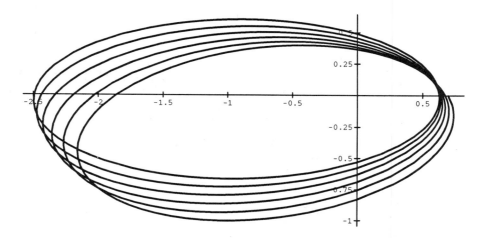

Figure 13.6: The advance of perihelion of planetary orbits

Advance of Perihelion of Planetary Orbits

It was a famous success of Einstein's general relativity theory in 1915 that it predicted the correct value for the advance of the perihelion (Greek for nearest the sun) of Mercury: 43 seconds of arc per century. Since that time it has correctly predicted the values for other planets and satelites. In general relativity, planetary orbits are approximated by geodesics (Greek for divides the Earth, because that is what great circles do on a sphere) which are curves of extremal length. The equations for the appropriate geodesics are rather involved but we know that they have approximate solutions that look like slowly rotating ellipses. The direction of the rotation of the ellipse is the same as that of the planet following the orbit; hence the reference to the advance of the perihelion. The approximate solution that comes from the theory is given in polar form by the radius function:

$$r = (1 + e\cos(\theta - k\theta))^{-1}$$

where e and k are parameters determined by the planet and the sun. The most convenient way to show the orbit is by using `ParametricPlot`.

For purposes of illustration we choose some values for k and e and assume that the ratio of major to minor axis of our planet is 0.5. Then we proceed as follows, renaming angle θ as a for convenience.

```
In[1]:= r[e_,k_,a_]:= 1/(1+e Cos[a-k a]);
In[2]:= e=0.6
In[3]:= k=0.02
In[4]:= ParametricPlot[{r[e,k,a] Cos[a],0.5 r[e,k,a] Sin[a]},
        {a,0, 2 Pi}, AspectRatio->Automatic];
```

This plots one complete orbit, but to see the advance of the perihelion we need to see several orbits in succession. This is easily done by the input:

```
In[5]:= ParametricPlot[{r[e,k,a] Cos[a],0.5 r[e,k,a] Sin[a]},
          {a,0, 12 Pi}, AspectRatio->Automatic];
```

The result is shown in Figure 13.6. You can experiment with larger families of orbits and try to fit the parameters for Mercury and the other planets, then check the periods of their orbits by computing arc lengths. The substitution $u = 1/r$ is helpful if you wish to see what differential equation is controlling the orbits; try differentiating u twice with respect to θ.

For further information:
C.T.J. Dodson and T. Poston. **Tensor Geometry** Graduate Texts in Mathematics 130. Springer-Verlag, New York 1991.

Appendix A

Mathematica Functions Introduced

A.1 Chap2.m

```
epsilondeltatest/:
  epsilondeltatest[f_, {a_, d_}, {l_, e_}] :=
    Block[{x}, Plot[f[x], {x, a - d, a + d},
PlotRange -> {l + e, l - e}]]

epsilondeltatest::usage = "epsilondeltatest[f, {a, d}, {l, e}]
plots a graph
of f on the interval {a - d, a + d} to see if |f[x] - l| < e for all x
such that |x - a| < d."

repsilondeltatest/:
  repsilondeltatest[f_, {a_, d_}, {l_, e_}] :=
    Block[{x}, Plot[f[x], {x, a, a + d},
PlotRange -> {l - e, l + e}]]
repsilondeltatest::usage = "repsilondeltatest[f, {a, d}, {l, e}]
 plots a graph
of f on the interval {a, a + d} to see if |f[x] - l| < e for all x
such that 0 < x - a < d."

lepsilondeltatest/:
  lepsilondeltatest[f_, {a_, d_}, {l_, e_}] :=
    Block[{x}, Plot[f[x], {x, a - d, a}, PlotRange -> {l - e, l + e}]]

lepsilondeltatest::usage = "lepsilondeltatest[f, {a, d}, {l, e}]
plots a graph
of f on the interval {a - d, a} to see if |f[x] - l| < e for all x
```

such that 0 < a - x < d."

A.2 Chap3.m

```
secant/: secant[f_, a_][p_] :=
(one = Block[{x}, Plot[{f[x], (f[p] - f[a]) (x - a)/(p - a) + f[a]},
      {x, a - 1.5, a + 1.5}, Axes -> None,
AspectRatio -> Automatic,
                    DisplayFunction -> Identity]];
    two = ListPlot[{{a, f[a]}, {p, f[p]}},
DisplayFunction -> Identity];
      Show[one, two, DisplayFunction -> $DisplayFunction])

secant::usage = "secant[f, a][p] plots f and the secant joining
(a, f[a]) and (p, f[p])."

tangent/: tangent[f_, a_][x_] := Derivative[1][f][a]*(x - a) + f[a]

normal/: normal[f_, a_][x_] := (a - x)/Derivative[1][f][a] + f[a]

tangentplot/: tangentplot[f_, a_, {b_, c_}] :=
   Plot[{f[x], tangent[f, a][x]}, {x, b, c}]

normalplot/: normalplot[f_, a_, {b_, c_}] :=
Plot[{f[x], normal[f, a][x]}, {x, b, c}, AspectRatio -> Automatic]

tangentfamily/:
  tangentfamily[f_, {a_, b_, stepsize_:0.1}] :=
   (one = Plot[f[x], {x, a, b}, DisplayFunction -> Identity];
    two =
    Block[{i}, Plot[Release[Table[tangent[f, i][x],
{i, a, b, stepsize}]],
     {x, a, b}, DisplayFunction -> Identity]];
      Show[one, two, DisplayFunction ->
$DisplayFunction])

tangentfamily::usage = "tangentfamily[f, {a, b, stepsize}]
plots f and its
tangents on the interval {a, b}, plotting one tangent per
stepsize in the interval {a, b}. The default stepsize is 0.1."

normalfamily/: normalfamily[f_, {a_, b_, stepsize_:0.1},
```

A.2. CHAP3.M

```
{c_, d_}] :=
   (one =
     Block[{i, x}, Plot[Release[Table[normal[f, i][x],
{i, a, b, stepsize}]], {x, a, b}, PlotRange -> {c, d},
AspectRatio -> Automatic,
     DisplayFunction -> Identity]];
        two = Block[{x}, Plot[f[x], {x, a, b},
DisplayFunction -> Identity]];
       Show[one, two,
DisplayFunction -> $DisplayFunction])

normalfamily::usage =
"normalfamily[f, {a, b, stepsize}, {c, d}]
plots f and its normals on the interval {a, b}, with PlotRange
set at {c, d}. One normal is plotted per stepsize in the interval
{a, b}; the defaultstepsize is 0.1."

wire/: wire[x_] :=
   (line = Block[{t}, Plot[70, {t, 0, 100},
 DisplayFunction -> Identity]];
      point = ListPlot[{{x, 70}}, DisplayFunction -> Identity];
      circle = Block[{t}, ParametricPlot[{(x*Cos[t])/(2*Pi) + 25,
        (x*Sin[t])/(2*Pi) + 25}, {t, 0, 2*Pi},
AspectRatio -> Automatic,
        DisplayFunction -> Identity]];
      square = ListPlot[{{75 - (100 - x)/8, 25 - (100 - x)/8},
        {75 - (100 - x)/8, 25 + (100 - x)/8},
        {75 + (100 - x)/8, 25 + (100 - x)/8},
        {75 + (100 - x)/8, 25 - (100 - x)/8},
        {75 - (100 - x)/8, 25 - (100 - x)/8}},
         PlotJoined -> True,  DisplayFunction -> Identity];
      Show[circle, square, line, point, Axes -> None,
         PlotRange -> {{0, 100}, {0, 80}}, PlotLabel ->
       N[x^2/(4 Pi) + (100 - x)^2/(16)] " = combined area of circle
and square",
        DisplayFunction -> $DisplayFunction])

ladder/: ladder[t_] :=
ListPlot[{{10 t, 0}, {0, Sqrt[41^2 - (10 t)^2]}},
         PlotRange -> {{0, 41}, {0, 41}},
         PlotJoined -> True, AspectRatio -> Automatic]
```

A.3 Chap4.m

```
bisect/: bisect[f_][{a_, b_}] :=
   If[N[f[a]*f[b]] > 0, {a, b},
   If[N[f[a]*f[(a + b)/2]] <= 0,
   N[{a, (a + b)/2}], N[{(a + b)/2, b}]]]

bisect::usage = "bisect[f][{a, b}] determines which half of the
interval {a, b} contains a solution to the equation f[x] == 0.
It is assumed that f is continuous on {a, b}.
If f[a] and f[b] have the same sign,
then the output is the original interval."

tangent/: tangent[f_, a_][x_] := f'[a]*(x - a) + f[a]
normal/: normal[f_, a_][x_] := (a - x)/f'[a] + f[a]

tangentplot/: tangentplot[f_, a_, {b_, c_}] :=
   Plot[{f[x], tangent[f, a][x]}, {x, b, c}]

normalplot/: normalplot[f_, a_, {b_, c_}] :=
   Plot[{f[x], normal[f, a][x]}, {x, b, c},
   AspectRatio -> Automatic]

errorplot/: errorplot[f_, a_, {b_, c_}] :=
   Plot[Abs[f[x] - tangent[f, a][x]], {x, b, c}, PlotRange -> All]

errorplot::usage = "errorplot[f, a, {b, c}]
plots the absolute value of
the difference of f and the tangent to f at x = a, for each x in the
interval {b, c}."

iteration/: iteration[f_][x_] := x - f[x]/f'[x]

newtonsmethod/:
   newtonsmethod[f_, a_, {b_, c_}][n_] :=
      (one =
Block[{x, i}, Plot[{f[x],
tangent[f, Nest[iteration[f], N[a], n]][x]},
{x,First[Union[{b,Table[Nest[iteration[f], N[a], i], {i,0,n+1}]]],
Last[Union[{c, Table[Nest[iteration[f], N[a], i], {i, 0, n + 1}]]]},
   DisplayFunction -> Identity]];
      two =
      Block[{i}, ListPlot[Table[{Nest[iteration[f], N[a], i], 0},
```

A.3. CHAP4.M

```
        {i, 0, n + 1}],
            DisplayFunction -> Identity]];
        three =
        Block[{i}, ListPlot[Table[{Nest[iteration[f], N[a], i],
            f[Nest[iteration[f], N[a], i]]}, {i, 0, n}],
                DisplayFunction -> Identity]];
        Show[one, two, three, PlotLabel ->
            If[n == 0, {"beginning Newton's Method",
            If[a == 0, "0 = initial guess", If[a == 1,
            "1 = initial guess", a " = initial guess"]]},
                {(n + 1) " iterations of Newton's method completed",
                    N[Nest[iteration[f], a, n + 1]] " = approximation"} ],
                DisplayFunction -> $DisplayFunction])

newtonsmethod::usage = "newtonsmethod[f, a, {b, c}][n]
shows Newton's method being used to generate the (n + 1)st
approximation to the solution of the equation f[x] == 0
from the nth approximation, starting with initial guess x0 = a.
All previous approximations are shown along the
x-axis, and the value of the current approximation is indicated
in the PlotLabel.
The interval {b, c} is the initial x-interval used for the
plot; however the x-interval will adjust automatically,
if necessary,
to include all approximations x0, x1, ... , xn, x(n + 1)."

fixedpoint/: fixedpoint[g_, a_, {b_, c_}][n_] :=
    (one =
    Block[{x, i}, Plot[{g[x], x},
    {x, First[Union[{b}, Table[Nest[g, N[a], i], {i, 0, n + 1}]]],
        Last[Union[{c}, Table[Nest[g, N[a], i], {i, 0, n + 1}]]]},
        AspectRatio -> Automatic, DisplayFunction -> Identity]];
    two =
    Block[{i}, ListPlot[Table[{Nest[g, N[a], i], 0}, {i, 0, n + 1}],
            DisplayFunction -> Identity]];
    three =
    Block[{i}, ListPlot[Table[{Nest[g, N[a], i], Nest[g, N[a], i]},
            {i, 0, n + 1}], DisplayFunction -> Identity]];
    four =
    Block[{i}, ListPlot[Table[{Nest[g, N[a], i],
    Nest[g, N[a], i + 1]},
            {i, 0, n + 1}], DisplayFunction -> Identity]];
    five =
```

```
        Block[{i}, ListPlot[Table[{Nest[g, N[a], Floor[i/2]],
            Mod[i + 1, 2]*Nest[g, N[a], Floor[i/2]] +
            Mod[i, 2]*g[Nest[g, N[a], Floor[i/2]]]}, {i, 0, 2 n + 3}],
            PlotJoined -> True, DisplayFunction -> Identity]];
        Show[one, two, three, four, five,
            PlotLabel -> If[n == 0,
                    {"beginning fixed point iteration",
            If[a == 0, "0 = initial guess", If[a == 1,
                "1 = initial guess", a " = initial guess"]]},
        {(n + 1) " iterations completed", N[Nest[g, a, n + 1]]
" = approximation"}],
                DisplayFunction -> $DisplayFunction])

fixedpoint::usage = "fixedpoint[f, a, {b, c}][n] performs (n + 1)
iterations of the Fixed Point iteration method for approximating
solutions to the equation f[x] == x, starting with x0 = a. The
output plots f and the identity function, y = x; shows all
successive approximations x0,
x1, ... , xn, x(n + 1) along the x-axis; indicates the value of the
current approximation in the PlotLabel; and shows the
'cobweb' of points homing in on the fixed point.
The initial x-interval is taken as {b, c},
but this will adjust automatically, if necessary, to include the
values of all approximations."
```

A.4 Chap5.m

```
curvature/: curvature[f_][x_] :=
Derivative[1][Derivative[1][f]][x]/(1 + Derivative[1][f][x]^2)^(3/2)

radiusofcurvature/:
radiusofcurvature[f_][x_] := Abs[1/curvature[f][x]]

viewcurvature/: viewcurvature[f_, {a_, b_}] :=
            Block[{x}, Plot[{f[x], curvature[f][x]}, {x, a, b},
        PlotLabel -> {f[x], "its graph and its curvature"}]]

viewcurvature::usage =
"viewcurvature[f, {a, b}] plots f and its
curvature function on the interval {a, b}."

centreofcurvature/: centreofcurvature[f_][s_] :=
```

A.4. CHAP5.M

```
      {s - (Derivative[1][f][s]*(1 + Derivative[1][f][s]^2))/
           Derivative[1][Derivative[1][f]][s],
        f[s] + (1 + Derivative[1][f][s]^2)/
                   Derivative[1][Derivative[1][f]][s]}

centreofcurvature::usage = "centreofcurvature[f][s]
finds the co-ordinates
of the centre of curvature of f at x = s."

viewevolute/: viewevolute[f_, {a_, b_}] := Block[{x, t},
        ParametricPlot[centreofcurvature[f][t], {t, a, b},
          PlotLabel -> {"the evolute of", f[x]}]]

viewevolute::usage = "viewevolute[f, {a, b}] plots
the centres of curvatue
of f for all x in the interval {a, b}. This curve is called the
evolute of f."

viewosculatingcircle/: viewosculatingcircle[f_][s_] :=
        (one = Block[{t, x}, ParametricPlot[
 {radiusofcurvature[f][s]*Cos[t] + centreofcurvature[f][s][[1]],
  radiusofcurvature[f][s]*Sin[t] + centreofcurvature[f][s][[2]]},
          {t, 0, 2*Pi}, AspectRatio -> Automatic,
             DisplayFunction -> Identity]];
         two = ListPlot[{centreofcurvature[f][s], {s, f[s]}},
             DisplayFunction -> Identity];
         three = Block[{x}, Plot[f[x],
       {x, centreofcurvature[f][s][[1]] - 3*radiusofcurvature[f][s],
           centreofcurvature[f][s][[1]] + 3*radiusofcurvature[f][s]},
        PlotRange ->
 {centreofcurvature[f][s][[2]] - 3*radiusofcurvature[f][s],
  centreofcurvature[f][s][[2]] + 3*radiusofcurvature[f][s]},
             DisplayFunction -> Identity]];
         Block[{x}, Show[one, two, three, PlotLabel ->
              {f[x], "its osculating circle at x =", s},
                DisplayFunction -> $DisplayFunction ]])

viewosculatingcircle::usage =
"viewosculatingcircle[f][s] plots the
osculating circle of f tangent to f at x = s."

survey/: survey[f_, {a_, b_}, s_] :=
     (one = Block[{t}, ParametricPlot[
```

```
           {radiusofcurvature[f][s]*Cos[t] + centreofcurvature[f][s][[1]],
            radiusofcurvature[f][s]*Sin[t] + centreofcurvature[f][s][[2]]},
               {t, 0, 2*Pi}, AspectRatio -> Automatic,
            DisplayFunction -> Identity]];
         two = Block[{v, u}, If[N[f'[s]] == 0,
         ParametricPlot[{s, f[s] + v},
            {v, -N[radiusofcurvature[f][s]],
             N[radiusofcurvature[f][s]]},
                   DisplayFunction -> Identity],
  Plot[(s - u)/
f'[s] + f[s], {u, s - N[Abs[s - centreofcurvature[f][s][[1]]]],
                 s + N[Abs[s - centreofcurvature[f][s][[1]]]]},
                   DisplayFunction -> Identity]]];
         three = Block[{m}, Plot[f'[s]*(m - s) + f[s], {m, s - 1, s + 1},
                   DisplayFunction -> Identity]];
         four = Block[{x}, Plot[{f[x], curvature[f][x]}, {x, a, b},
                   DisplayFunction -> Identity]];
         five = Block[{z}, ParametricPlot[centreofcurvature[f][z], {z, a, b},
                   DisplayFunction -> Identity]];
         six = ListPlot[{{s, f[s]}, centreofcurvature[f][s]},
                   DisplayFunction -> Identity];
         Block[{x}, Show[one, two, three, four, five, six,
           PlotLabel -> {f[x], "survey of curvature details at x =", s},
                   DisplayFunction -> $DisplayFunction ]])

survey::usage =
"survey[f, {a, b}, s] plots f, its curvature function
and its evolute for x in the interval {a, b}.
In addition, it includes the osculating circle tangent to f at
x = s, plus short normal and tangent lines to f at x = s."

d/: d[p_, q_] := 4*p^3 + 27*q^2

cubicplot/: cubicplot[p_, q_] := Block[{x}, If[p < 0,
Plot[x^3 + p*x + q,
{x, N[-(3*Abs[Sqrt[-p]/Sqrt[3]])], N[3*Abs[Sqrt[-p]/Sqrt[3]]]},
     PlotLabel -> Which[d[p, q] > 0,
{x^3 + p x + q, "one real root", If[d[p, q] == 1, "1 = D",
                                                 d[p, q] " = D"]},
                    d[p, q] == 0,
{x^3 + p x + q, "two real roots", "0 = D"},
                    d[p, q] < 0,
{x^3 + p x + q, "three real roots", If[d[p, q] == -1, "-1 = D",
```

```
                                              d[p, q] " = D"]}]],
     Plot[x^3 + p*x + q, {x, N[-3*q^(1/3)] - 1, N[3*q^(1/3)] + 1},
                    PlotLabel ->
     {x^3 + p x + q, "one real root", " p is non-negative"}]]]

cubicplot::usage =
"cubicplot[p, q] plots the cubic x^3 + p*x + q
on a suitably chosen interval.
The PlotLabel indicates how many real
solutions there are to the equation x^3 + p*x + q == 0,
and also indicates the value
of the discriminant, D = 4*p^3 + 27*q^2."
```

A.5 Chap6.m

```
anti/: anti[tableofvalues_][1] = 0

anti/: anti[tableofvalues_][n_] :=
   anti[tableofvalues][n] =
    anti[tableofvalues][n - 1] +
((First[Transpose[tableofvalues]][[n]] -
         First[Transpose[tableofvalues]][[n - 1]])*
       (Last[Transpose[tableofvalues]][[n]] +
         Last[Transpose[tableofvalues]][[n - 1]]))/2

antiderivative/: antiderivative[tableofvalues_, k_] :=
Block[{i},
   Table[{First[Transpose[tableofvalues]][[i]],
anti[tableofvalues][i] + k},
     {i, Length[tableofvalues]}]]

viewantiderivative/:
viewantiderivative[f_, {a_, b_}, initialvalue_] :=
    (one = Block[{x}, Plot[f[x], {x, a, b},
 DisplayFunction -> Identity]];
     two = Nest[First, InputForm[one], 5];
     Block[{x},
ListPlot[antiderivative[two, initialvalue], PlotJoined -> True,
    PlotLabel -> {"an approximate antiderivative of", f[x]}]])

viewantiderivative::usage =
"viewantiderivative[f, {a, b}, initialvalue]
```

plots the graph of an approximate antiderivative F of f
on the interval {a,b} with initial condition F[a] = initialvalue.
The approximation is based
on applying the Mean Value Theorem to Mathematica's
InputForm of the Plot
of f on {a, b}."
antiderivativefamily/:
  antiderivativefamily[f_, {a_, b_, k_:1}] :=
(one = Block[{x}, Plot[f[x], {x, a, b}, AspectRatio -> Automatic,
    DisplayFunction -> Identity]];
      two = Block[{x}, Integrate[f[x], x] /. x -> a];
  three = Block[{x, i}, Plot[Release[Table[Integrate[f[x], x] + i,
        {i, two - 5, two + 5, k}]], {x, a, b},
              DisplayFunction -> Identity]];
      Block[{x}, Show[one, three,
PlotLabel -> {"antiderivatives of", f[x]},
      DisplayFunction -> $DisplayFunction]])

antiderivativefamily::usage = "antiderivativefamily[f, {a, b, k}]
plots F[x] + c, and f[x], on the interval {a, b} where
F[x] = Integrate[f[x], x], for a few values of c differing by k.
The default value for k is 1."

A.6 Chap7.m

riemann/: riemann[f_, {a_, b_}, n_] :=
  N[(b - a)*Sum[f[a + i*(b - a)/n], {i, 1, n}]/n]

riemann::usage = "riemann[f, {a, b}, n] calculates the
numerical value of the Riemann sum obtained by dividing
the interval {a, b} into n equal
subintervals and evaluating f at the right hand end
point of each subinterval."

er/: er[f_, {a_, b_}, n_] :=
Abs[NIntegrate[f[x], {x, a, b}] - riemann[f, {a, b}, n]]

er::usage = "er[f, {a, b}, n] computes the absolute value of the
difference between the value of the definite integral of f
on the interval {a, b} and the value of the Riemann
sum approximation riemann[f, {a, b}, n]."

A.6. CHAP7.M

```
definitetransform/:
definitetransform[integrand_, {x_, a_, b_}, u_, substitution_] :=
{newintegrand = integrand/D[substitution, x] /. substitution -> u,
u == substitution, {substitution /. x -> a, substitution /. x -> b}}

definitetransform::usage =
"definitetransform[integrand, {x, a, b}, u, substitution]
applies the change of variables u = substitution (in terms
of x) to the definite integral of the given integrand (in
terms of x) on the interval {a, b} and outputs a list
with three elements: the first element is the new integrand
(newintegrand) in terms of u (and possibly still x);
the second element is an equation giving u in terms of
x; the third element is a list of replacements of the new limits of
integration in terms of u."
s/: s[f_, {a_, b_}, x_] := If[a <= x <= b, f[b], 0]

randompartition/: randompartition[{a_, b_}, n_] := P =
    Union[Table[Random[Real, {N[a], N[b]}], {n - 1}], {N[a], N[b]}]

randompartition::usage = "randompartition[{a, b}, n]
produces a random partition of the interval {a, b}
into at most n subintervals by randomly
picking n - 1 points in {a, b}. The partition is set equal to P."
regularpartition/: regularpartition[{a_, b_}, n_] := P = Block[{i},
    Table[N[a + i (b - a)/n], {i, 0, n}]]
regularpartition::usage = "regularpartition[{a, b}, n] partitions
the interval {a, b} into n equally spaced subintervals.
The partition is set equal to P."

refine/: refine[partition_, n_] := P = Union[Table[Random[Real,
{Min[N[partition]], Max[N[partition]]}], {n}], N[partition]]

refine::usage = "refine[partition, n] randomly inserts n
(possibly not distinct) points into the given partition.
The new partition is set equal to P."

arbitraryriemann/:
    arbitraryriemann[f_, P_] := Block[{i},
Sum[(Union[N[P]][[i + 1]] -
Union[N[P]][[i]])*f[Union[N[P]][[i + 1]]],
    {i, 1, Length[P] - 1}]]
arbitraryriemann::usage = "arbitraryriemann[f, P]
```

calculates the Riemann sum of f for the partition P by evaluating
f at the right end point of each subinterval of P."

ear/: ear[f_, P_] := Block[{x}, Abs[NIntegrate[f[x],
{x, Min[N[P]], Max[N[P]]}] - arbitraryriemann[f, P]]]

ear::usage = "ear[f, P] calculates the difference between
the value of the defintite integral of f on the interval
{Min[P], Max[P]} and arbitraryriemann[f, P]."

norm/: norm[P_] :=
Max[Drop[Union[N[P]], 1] - Drop[Union[N[P]], -1]]
norm::usage = "norm[P] is the norm of the partition P."

viewapprox/: viewapprox[f_, P_] :=
(one = Block[{x}, Plot[f[x], {x, Min[N[P]] - 0.1, Max[N[P]] + 0.1},
        DisplayFunction -> Identity]];
   two = Block[{i, x},
Plot[Release[Table[s[f, {Union[N[P]][[i]], Union[N[P]][[i + 1]]}, x],
        {i, Length[P] - 1}]], {x, Min[N[P]] - 0.1, Max[N[P]] + 0.1},
PlotRange -> All,
PlotPoints -> 50, DisplayFunction -> Identity]];
     Block[{x}, Show[one, two, PlotLabel ->
        {f[x], (Length[P] - 1)*"intervals" , N[norm[P]]*"= norm",
  If[Chop[ear[f, P]] == 0, "no error", ear[f, P]*"= error"]},
     PlotRange -> All, DisplayFunction -> $DisplayFunction]])

viewapprox::usage = "viewapprox[f, P] shows the Riemann sum
approximation to the definite integral of f on the interval
{Min[P], Max[P]} in terms of the given partition P.
The PlotLabel includes the function f, the number
of subintervals, the norm of the partition, and the absolute
error of the Riemann approximation.
If P is a partition with some points extremely
close together - as may happen with a random partition -
the narrow rectangles on these short subintervals may
be missed in the output."

average/: average[f_, {a_, b_}] := Integrate[f[x], {x, a, b}]/(b - a)

average::usage = "average[f, {a, b}] is the average value of f on the
interval {a, b}."

A.7 Chap8.m

```
riemann/: riemann[f_, {a_, b_}, n_] :=
    (b - a)*Sum[f[a + i*(b - a)/n], {i, 1, n}]/n // N
riemann::usage = "riemann[f, {a, b}, n] calculates the numerical
value of the Riemann sum obtained by dividing the interval {a, b}
into n equally spaced subintervals and evaluating f at the right
hand end point of each subinterval."

trapezoid/: trapezoid[f_, {a_, b_}, n_] :=
  ((b - a)*(f[a] + 2*Sum[f[a + (i*(b - a))/n], {i, 1, n - 1}] +
            f[b]))/(2*n) // N
trapezoid::usage =
"trapezoid[f, {a, b}, n] calculates the trapezoid rule
approximation of the definite integral of f on the interval {a, b} by
dividing the interval {a, b} into n equally spaced subintervals."

simpson/: simpson[f_, {a_, b_}, n_] :=
   ((b - a)*(f[a] + 4*Sum[f[a + (i*(b - a))/n], {i, 1, n - 1, 2}] +
  2*Sum[f[a + (j*(b - a))/n], {j, 2, n - 2, 2}] + f[b]))/(3*n) // N

simpson::usage =
"simpson[f, {a, b}, n] calculates the Simpson's rule
approximation of the definite integral of f on the interval {a, b} by
dividing the interval {a, b} into n equally spaced subintervals,
where n is even."

er/: er[f_, {a_, b_}, n_] :=
   Abs[NIntegrate[f[x], {x, a, b}] - riemann[f, {a, b}, n]]
er::usage = "er[f, {a, b}, n] computes the absolute value of the
difference between the value of the definite integral of f on the
interval {a, b} and the value of the Riemann sum approximation
riemann[f, {a, b}, n]."

et/: et[f_, {a_, b_}, n_] :=
   Abs[NIntegrate[f[x], {x, a, b}] - trapezoid[f, {a, b}, n]]

et::usage = "et[f, {a, b}, n] computes the absolute value of the
difference between the value of the definite integral of f on the
interval {a, b} and the value of the trapezoid rule approximation
trapdezoid[f, {a,b}, n]."

es/: es[f_, {a_, b_}, n_] :=
```

```
      Abs[NIntegrate[f[x], {x, a, b}] - simpson[f, {a, b}, n]]

es::usage = "es[f, {a, b}, n] computes the absolute value of the
difference between the value of the definite integral of f on the
interval {a, b} and the value of the Simpson's rule approximation
simpson[f, {a,b}, n]."

s/: s[f_, {a_, b_}, x_] := If[a <= x <= b, f[b], 0]


q/: q[f_, {a_, b_}, x_] :=
    If[a <= x <= b, f[a] + (2*(f[(a + b)/2] - f[a])*(x - a))/(b - a) +
  (2*((f[a] + f[b]) - 2*f[(a + b)/2])*(x - a)*(x - (a + b)/2))/
              (b - a)^2, 0]

l/: l[f_, {a_, b_}, x_] :=
    If[a <= x <= b, ((f[b] - f[a])*(x - a))/(b - a) + f[a], 0]

viewrapprox/: viewrapprox[f_, {a_, b_}, n_] :=
 (one = Block[{x}, Plot[f[x], {x, a, b},
   DisplayFunction -> Identity]];
      two = Block[{i, x},
Plot[Release[Table[s[f, {a + ((i - 1)*(b - a))/n,
a + (i*(b - a))/n}, x],
        {i, n}]], {x, a, b}, PlotRange -> All, PlotPoints -> 50,
             DisplayFunction -> Identity]];
         Block[{x}, Show[one, two, PlotLabel ->
       {f[x], n*"intervals", N[(b - a)/n]*"= norm",
If[Chop[er[f, {a, b}, n]] == 0, "no error",
er[f, {a, b}, n]*"= error"]},
  PlotRange -> All,
  DisplayFunction -> $DisplayFunction ]])

viewrapprox::usage =
"viewrapprox[f, {a, b}, n] shows the Riemann
sum approximation to the definite integral of f on the interval
{a, b} for n equally spaced subintervals of the given interval
{a, b}. The PlotLabel includes the function f, the number of
subintervals, the norm of the partition, and the absolute error
of the Riemann approximation."

viewtapprox/: viewtapprox[f_, {a_, b_}, n_] :=
 (one = Block[{x}, Plot[f[x],
```

A.7. CHAP8.M

```
    {x, a, b}, DisplayFunction -> Identity]];
       two = Block[{x, i}, Plot[Release[Table[
    l[f, {a + ((i - 1)*(b - a))/n, a + (i*(b - a))/n}, x],
       {i, n}]], {x, a, b}, PlotRange -> All, PlotPoints -> 50,
          DisplayFunction -> Identity]];
       Block[{x}, Show[one, two, PlotLabel ->
       {f[x], n*"intervals", N[(b - a)/n]*"= norm",
If[Chop[et[f, {a, b}, n]] == 0, "no error",
et[f, {a, b}, n]*"= error"]}, PlotRange -> All,
DisplayFunction -> $DisplayFunction ]])

viewtapprox::usage =
"viewtapprox[f, {a, b}, n] shows the trapezoid
rule approximation to the definite integral of f on
the interval {a, b}
for n equally spaced subintervals of the given interval {a, b}.
The PlotLabel includes the function f, the number of subintervals,
the norm of the partition, and the absolute error of the
trapezoid rule approximation."

viewsapprox/: viewsapprox[f_, {a_, b_}, n_] :=
  (one = Block[{x}, Plot[f[x], {x, a, b},
DisplayFunction -> Identity]]; two = Block[{x, i},
       Plot[Release[Table[
    q[f, {a + ((i - 1)*(b - a))/n, a + ((i + 1)*(b - a))/n}, x],
       {i, 1, n - 1, 2}]], {x, a, b},
       PlotRange -> All, PlotPoints -> 50,
          DisplayFunction -> Identity]];
       Block[{x}, Show[one, two, PlotLabel ->
       {f[x], n*"intervals", N[(b - a)/n]*"= norm",
    If[Chop[es[f, {a, b}, n]] == 0, "no error",
es[f, {a, b}, n]*"= error"]},
PlotRange -> All, DisplayFunction -> $DisplayFunction]])

viewsapprox::usage = "viewsapprox[f, {a, b}, n]
shows the Simpson's rule approximation to the definite integral of f
on the interval {a, b} for n equally spaced subintervals of the given
interval {a, b}, where n must be even.
The PlotLabel includes the function f, the number of subintervals,
the norm of the partition, and the absolute error of the
Simpson's rule approximation."
```

A.8 Chap11.m

```
in[n_]:= n/2(Sin[2 Pi/n])   (* Inscribed n-gon area *)
ex[n_]:= n(Tan[Pi/n])       (* Exscribed n-gon area *)
ListPlot[Table[n/2(Sin[2 Pi/n]) ,{n,4,100}],
                AxesLabel->{"n",""}];
ListPlot[Table[n(Tan[Pi/n]) ,{n,4,100}],
                AxesLabel->{"n",""}];
```

A.9 Chap12.m

```
parts/: parts[integrand_, x_, u_, substitution_] :=
{newintegrand =
Integrate[integrand/substitution, x]*D[substitution, x],
 u -> substitution, v -> Integrate[integrand/substitution, x]}

parts::usage =
"parts[integrand, x, u, substitution] applies integration
by parts to the original integrand (in terms of x)
with u = substitution.
It outputs a list of three elements:
the first member is the integrand u
dv (newintegrand); the second indicates the choice of u;
the third indicates the value of v."

transform/: transform[integrand_, x_, u_, substitution_] :=
    {newintegrand =
integrand/D[substitution, x] /. substitution -> u,
    u == substitution}

transform::usage =
"transform[integrand, x, u, substitution] applies the
change of variables u = substituion (in terms of x) to the original
integrand (in terms of x). It outputs a list of two elements:
the first element is the new integrand (newintegrand)
in terms of u (and possibly
x); the second element is an equation giving the substitution u in
terms of x."

inversesub/: inversesub[integrand_, x_, z_, substitution_] :=
    {newintegrand =
```

A.10 Chap13.m

```
       (integrand /. x -> substitution)*D[substitution, z],
          Solve[x == substitution, z]}

inversesub::usage =
"inversesub[integrand, x, z, substitution] applies the
change of variables x = substitution (in terms of z) to the original
integrand (in terms of x). It outputs a list of two elements:
the first element is the new integrand (newintegrand) in terms of z;
 the second element is a list of solutions to the equation
 x == substitution,  giving z as one or more functions of x."

completesquare/: completesquare[quadratic_, x_] :=
       Coefficient[quadratic, x^2]*
       (x + Coefficient[quadratic, x]/
 (2*Coefficient[quadratic, x^2]))^2 +
       Coefficient[quadratic, x^2]*
       ((4*Coefficient[quadratic, x^2]*(quadratic /. x -> 0) -
             Coefficient[quadratic, x]^2)/
 (4*Coefficient[quadratic, x^2]^2))

completesquare::usage =
"completesquare[quadratic, x] completes the square
of the quadratic function of x."
```

A.10 Chap13.m

```
points/: points[{a_, b_}, {c_, d_}] := Block[{x, y},
   Table[N[{x, y}], {x, a, b, (b - a)/10}, {y, c, d, (d - c)/10}]]

attractors/: attractors[f_, {a_, b_}, {c_, d_}, n_] :=
   Block[{x, y}, Table[Nest[iteration[f], N[x + y*I], n],
      {x, a, b, (b - a)/10}, {y, c, d, (d - c)/10}]]

f/: f[x_] := x^4 - 1

iteration/: iteration[f_][x_] := x - f[x]/Derivative[1][f][x]

colour/: colour[i_, j_] := Which[
          attractor[[i, j]] == 1.,   RGBColor[1, 0, 0],
          attractor[[i, j]] == 1. I, RGBColor[0, 1, 0],
          attractor[[i, j]] == -1.,  RGBColor[0, 0, 1],
          attractor[[i, j]] == -1. I, RGBColor[1, 1, 0],
```

```
          True, RGBColor[1, 0, 1]]]

patch/: patch[f_, {a_, b_}, {c_, d_}, n_] := (
        lattice = points[{a, b}, {c, d}];
        attractor = attractors[f, {a, b}, {c, d}, n];
        Show[Table[Graphics[{colour[i, j], Point[lattice[[i, j]]]}],
                   {i, 1, 11}, {j, 1, 11}]])

f[n_,t_]:=(1+100 E^(-t/2))^(-1/n)  (* Growth function *)
Plot[{f[1,t],f[2,t],f[3,t],f[4,t],f[5,t],f[6,t]},{t,0,20}];

r[e_,k_,a_]:= 1/(1+e Cos[a-k a]);  (* Orbit radius function *)
ParametricPlot[{r[e,k,a] Cos[a],0.5 r[e,k,a] Sin[a]}, {a,0,12 Pi},
AspectRatio->Automatic];
```

Bibliography

[1] D. Brown, H. Porta and J. Uhl. **Calculus and Mathematica** Addison-Wesley, California 1990. (See also their article: Calculus and Mathematica: Courseware for the Nineties. *Mathematica Journal* 1, 1, (1990) 43-50.)

[2] R.E. Crandall. **Mathematica for the Sciences** Addison-Wesley, California 1991.

[3] C.H. Edwards, Jr. and D.E. Penney. **Calculus and Analytic Geometry** 3rd edition, Prentice Hall, New Jersey 1990.

[4] T.W. Gray and J. Glynn. **Exploring Mathematics with Mathematica** Addison-Wesley, California 1991.

[5] L. Lamport. LaTeX **A Document Preparation System** Addison-Wesley, California 1986.

[6] R. Maeder. **Programming in Mathematica** Addison-Wesley, California 1990.

[7] D.W. Trim. **Single-Variable Calculus**, Prentice-Hall Canada Inc., Scarborough 1992.

[8] S. Wolfram. **Mathematica: A System of Doing Mathematics by Computer**, Addison-Wesley, California 1988.

Index

A
acceleration 210
acceleration due to gravity 156
adaptive numerical integration 205
Animate 121
animation 37, 97, 127, 136, 171
anti 166
antiderivative 151, 166
antiderivativefamily 156
applications of integrals 209
approximate antiderivative 164
arc length 209, 213
archaic units 101
area 209
AspectRatio 29, 86
Automatic 29
average 191

B
best linear approximation 67
best polynomial fit 201
bisect 110, 129
bisection method 109
brackets 15
built-in functions 6

C
calculation of areas 180
cardioid 216
catenary 234
catenoid 234
centreofcurvature 143, 149
chain rule 73, 153
change of variable 188
chaos 127
chocolate donut 212

Clear 12
clear function 8
ClickKillsWindows 1
closest points 90
concave 133, 142
cone 211
constructing antiderivatives 162
continuous 54
critical point 80, 94, 132
critical values 27
cross sections 210
cubic 83, 132, 137
cubicapprox 203
curvature 138
cylindrical shells 213

D
D 8
data points 29
decimal 3, 6
default precision 112, 121
definite integral 175, 177
definitetransform 188
Derivative 78
derivative 8, 69
derivatives of trigonometric functions 231
discontinuity 80, 155
discontinuous 54
Display 33
division 3
DOS to UNIX 2
dropped ball problem 160

E
E 4, 224

INDEX

e as a limit 225
editor 1
electronic mail 2
ellipse 34, 93, 149
ellipse parametrically 215
ellipsoid 213
encapsulated PostScript 33
envelope 86
envelope of normals 150
epsilondeltatest 60
epsilontics 60
er 177, 194
Erf 154
error expressions 194
error function 154
errorplot 105, 108
error term 177
evaluate 8
evolute 144, 150
exp 222
Expand 10
exponential decay 226
exponential function 223
exponential growth 226
exponentiation 3
expression names 12
extraction from lists 18
extreme values 94, 100

F
file transfer 2
FindRoot 120
Fit 29, 193
fixedpoint 124
FixedPoint 127
Fixed Point Iteration 123
Floor 49, 76, 190
flow rate 210
force 210
function 6
Fundamental Theorem of Calculus 136, 183, 193, 222

G

geometrical antiderivatives 188
geometric series 51
GIGO 3
graphing 24
greatest integer function 76
grouping characters 3

H
higher derivatives 78
hyperbola 81, 93, 234
hyperbolic functions 229, 233

I
I 4
increasing 133
indefinite integral 151
implicit differentiation 77
inflection point 132, 145
information 6
input 2, 3
inputting packages 36
integral 8
Integrate 151
Intermediate Value Property 108
intersection points 191
interrupt 4
inverse to logarithm 222
inverse trigonometric functions 232
iteration 127
iterator 16, 77

L
ladder 97
ladder problem 97
last output 4
L'Hôpital's Rule 136
limacon of Pascal 217
Limit 41
limit 41, 59
linear approximation 103
list 16
listable 105
listable functions 16
ListPlot 29

lituus 216
logarithm as integral 221

M
math 3
maximum curvature 147
Mean Value Theorem 131, 192
memory 2
Method of Bisection 108
minimum distance 89
Möbius strip 214
MoebiusStrip 214
more information 6
movie 121
multiplication 3
MVT 131

N
N 3
natural logarithm 5, 221
Needs 36, 37
Needs-alternative 36
Nest 111, 124
network 2
Newton fails 119
Newton's Method 114, 120
newtonsmethod 115
NIntegrate 183, 204
norm 176
normal 70, 80
norm of partition 176
normalfamily 86
normalplot 70, 84
NRoots 112
NSum 52
numerical integration comparison 199
numerical methods 193

O
one-sided limit 47
orthogonal 87
osculating circle 143
output 4
output number 4

P
package 36
ParametricPlot 27, 149, 214
ParametricPlot3D 217
parabola 84, 86, 87, 92, 190
partial derivative 79
pathname 36
partition 175
perpendicular 87
Pi 4
Plot 24
Plot3D 33, 212
points of intersection 89
Poisson distribution 227
pole problem 101
power series 76
printing problem 101
printers 2
properties of exponential 224
properties of logarithm 223
psfix 33
pumping fluid 211

Q
quartic 92
Quit 4
quotient rule 72

R
radius of curvature 142
radiusofcurvature 142
random processes 227
rate of convergence 128
rational function 134
reciprocal spiral 215
recycling exponential function 233
Reduce 22
refine partition 176
regular partition 175, 193
regularpartition 176
related rates 97
remarks in inputs 8
riemann 177, 194
Riemann sum 175

INDEX

Rolle's Theorem 136
root approximations 105
roots 123
rotating curves 217

S
saving graphics 33
secant 67
separable differential equations 218
shell escape 2
Show 32
Simplify 36, 152
simpson 194
Simpson's Rule 193
slope 69
solid of revolution 212
Solve 19, 113
sphere 210
spiral of Archimedes 215
substitution 12, 188
Sum 76
surface area of revolution 215
survey 146
symmetries 210

T
Table 16, 77
tableofvalues 166
tangent 67, 80, 103
tangentfamily 84
tangentplot 70, 84, 103
Three Eighths Rule 204, 206
thrown ball problem 159
torus 212
transcendental functions 229
trapezoid 194
Trapezoid Rule 193, 201
Trigonometry 36
trigonometric functions 229
trigonometric identities 13
TrigReduce 36, 38

U
UNIX 1

usage 7

V
variable 78
velocity 210
vertical asymptotes 55
vi editor 38
viewapprox 178
viewantiderivative 172
viewevolute 144
viewcurvature 141
viewosculatingcircle 144
viewsapprox 197, 206
viewtapprox 195
volume 209

W
wild-card 6
windows 1
wire 96
wire problem 95
wrong answer 41